Unity3D平台
AR开发快速上手
基于EasyAR 4.0

吴雁涛　赵　杰　叶东海　编著

清华大学出版社
北京

内 容 简 介

EasyAR 是国内很常用的一款免费增强现实引擎,也是国内 AR 开发常用的工具,具有使用简单、容易上手的特点,非常利于初学者学习使用。2019 年年底推出的 EasyAR 4.0 版本还提供了运动跟踪、稀疏空间地图和稠密空间地图等新功能,将原来针对平面内容进行识别的功能扩展到了三维现实空间。

本书共分为 10 章,内容包括增强现实开发基本知识、EasyAR 4.0 开发基础、平面图像跟踪相关功能、3D 物体跟踪、表面跟踪和运动跟踪、环境认知、EasyAR 学习使用小示例以及室内导航的实战案例,可以让读者掌握 EasyAR 4.0 开发 AR 应用的方法和技巧。

本书适合 Unity3D 平台 AR 开发初学者、手机游戏开发人员阅读,也适合作为高等院校和培训机构 AR 开发课程的教学参考书。

本书封面贴有清华大学出版社防伪标签,无标签者不得销售。
版权所有,侵权必究。举报:010-62782989,beiqinquan@tup.tsinghua.edu.cn。

图书在版编目(CIP)数据

Unity3D 平台 AR 开发快速上手:基于 EasyAR 4.0/吴雁涛,赵杰,叶东海编著. —北京:清华大学出版社,2021.1

ISBN 978-7-302-56931-2

Ⅰ. ①U… Ⅱ. ①吴… ②赵… ③叶… Ⅲ. ①游戏程序-程序设计 Ⅳ. ①TP317.6

中国版本图书馆 CIP 数据核字(2020)第 228140 号

责任编辑:夏毓彦
封面设计:王 翔
责任校对:闫秀华
责任印制:丛怀宇

出版发行:清华大学出版社
网　　址:http://www.tup.com.cn,http://www.wqbook.com
地　　址:北京清华大学学研大厦 A 座　　　　邮　编:100084
社 总 机:010-62770175　　　　　　　　　　邮　购:010-62786544
投稿与读者服务:010-62776969,c-service@tup.tsinghua.edu.cn
质量反馈:010-62772015,zhiliang@tup.tsinghua.edu.cn

印 装 者:三河市君旺印务有限公司
经　　销:全国新华书店
开　　本:190mm×260mm　　　　印　张:20　　　字　数:512 千字
版　　次:2021 年 1 月第 1 版　　　　　　　　印　次:2021 年 1 月第 1 次印刷
定　　价:79.00 元

产品编号:088928-01

推 荐 序

时值金秋，EasyAR迈入了第六个年头。在过去的五年间，EasyAR有幸从一个初出茅庐的SDK成长为国内外众多开发者首选的AR开发平台。

虽然AR在国内进入公众视野的时间并不长，但是AR技术演化已经有几十年的历史，身处这样一个技术日新月异的时代，我们是幸运的。不知从什么时候起，融入AR元素的科幻作品越来越多，像头号玩家、黑镜、刀剑神域等都有不小的热度。在被这些近未来的场景吊足胃口的同时，我们有时候真的快分不清现实与科幻的界限了，生怕一转身就被时代所抛弃。

为了能让虚拟形象走入现实，让物理世界增添更多便利，世界上有众多科技公司正在努力着，我们也是其中的一员。透过屏幕，EasyAR可以让虚拟偶像走进客厅，可以在路面显示导航路径，可以让你的照片播放视频，也可以在景区增置更多历史内容的动态展示。智能手机的发展让这些场景可以走入每个人的手中，随着5G技术的发展，更大规模的应用正在成为可能。

EasyAR的内部是各种图像以及多传感器融合算法：让照片和图片上出现视频或模型的，是平面图像的识别与跟踪；让地球仪表面出现动态效果的，是3D物体的跟踪；让模型始终贴附在运动表面上的，是融合了惯性数据的表面跟踪；让虚拟偶像像真人一样出现在现实世界中的，是基于视觉惯性里程计的运动跟踪；让虚拟偶像偶尔调皮地藏在桌子底下的，是实时生成的稠密空间地图。

导航和景区的内容展示随着规模的变化用到的技术可能有所不同。运动跟踪维持部分行进过程中的持续跟踪，稀疏空间地图让指定区域的环境被记住并能随时被找到，跟随稀疏空间地图存储和加载的稠密空间地图让虚拟物体在现实环境中的遮挡和交互成为可能。在场景规模从一个房间扩展到一个建筑、一座城市，乃至整个地球的时候，云端的定位技术就会被用到，在其背后则是一套完整的高精建图、定位和数据服务。

为了实现上述场景，EasyAR Sense SDK及其Unity插件将复杂的算法过程隐藏，通过简单易用的接口向开发者提供算法能力。同时，组件化的设计最大化了SDK的灵活性，使得自定义的相机输入、自定义的渲染流程乃至自定义的算法都可以在整个框架中运作。

Unity3D 是 AR 创作过程中常常被使用的 3D 引擎与工具。本书介绍了 EasyAR Sense SDK 4.0 在 Unity3D 平台下的使用，并以室内导航为例，从实战角度解析了一个 AR 项目的制作和调试发布过程，非常适合想要入门 AR 开发以及有一定 AR 开发经验的读者参考。

AR 这个领域还在不断发展，市面上已经有了很多读物，虽然有些还不太完善，但是 AR 导航已经可以使用了。随着新基建、5G 的推进，AR 的应用必将层出不穷。我们将继续站在行业前端，推动行业不断发展。

EasyAR 研发总监

宋　健

2020 年 12 月

前　　言

　　EasyAR 是国内很常用的一款免费增强现实引擎，也是国内 AR 开发常用的工具，具有容易上手，使用简单的特点，非常利于初学者学习使用。2019 年 12 月推出的 EasyAR 4.0 还提供了运动跟踪、稀疏空间地图和稠密空间地图等新功能，将原来针对平面内容进行识别的功能扩展到了三维现实空间。这样的进步使得 EasyAR 4.0 具备了以往没有的很多功能，最典型的就是基于增强现实的室内导航。

　　本书内容包括增强现实开发基本知识、EasyAR 4.0 开发基础、平面图像跟踪相关功能、3D 物体跟踪、表面跟踪和运动跟踪、环境认知、EasyAR 学习使用小示例以及室内导航的实战案例，可以让读者掌握 EasyAR 4.0 开发 AR 应用的方法和技巧。

本书特点

　　（1）快速上手：以最直接、最细致的方式指导读者快速掌握 EasyAR 4.0 的安装和使用。

　　（2）理解架构：通过例子和讲解，理解 EasyAR 4.0 能实现的功能及其实现方法。

　　（3）实战引导：通过实际项目，介绍 EasyAR 4.0 的具体实现，引导读者快速入门。

　　（4）热点跟进：室内导航的例子，让读者能够快速跟上 AR 应用开发的潮流。

示例源代码及其资源下载

　　本书配套的示例源代码及其资源下载，请扫描下面的二维码获取。如果有任何问题，请直接发邮件至 booksaga@163.com，邮件主题为"AR 开发快速上手"。

本书读者

本书适合 Unity3D 平台 AR 开发初学者、手机游戏开发人员阅读，也适合作为高等院校和培训机构计算机游戏开发课程的教学参考书。

本书作者

吴雁涛，2000 年毕业于西北工业大学材料科学与工程专业，从事计算机相关工作，包括网站建设、Web 前端、Unity3D 开发等。著有图书《Unity3D 平台 AR 与 VR 开发快速上手》《Unity 2018 AR 与 VR 开发快速上手》。吴雁涛是 EasyAR 官方认定的推广大使。

吴雁涛
2020 年 12 月

目　　录

第 1 章　EasyAR 4.0 基础 ···1
1.1　EasyAR 4.0 简介 ···1
1.2　下载导入和基本设置 ···2
1.2.1　获取 Key ···2
1.2.2　下载导入 ···5
1.2.3　基本设置 ···6
1.3　EasyAR 的基本结构 ··12
1.3.1　EasyAR 游戏对象下的设置 ··13
1.3.2　VideoCameraDevice 游戏对象下的设置 ··13
1.3.3　VIOCameraDevice 游戏对象下的设置 ··15

第 2 章　图像和物体跟踪 ··16
2.1　平面图像跟踪 ··16
2.1.1　总体说明 ···16
2.1.2　跟踪单个图像 ··18
2.1.3　跟踪多个图像 ··22
2.1.4　平面图像跟踪程序控制 ··25
2.2　平面图像跟踪扩展内容 ···29
2.2.1　视频播放 ···29
2.2.2　涂涂乐 ··33
2.3　云识别 ··37
2.3.1　总体说明 ···37
2.3.2　上传图片 ···37
2.3.3　添加基本内容 ··39
2.3.4　相关程序控制 ··40
2.4　3D 物体跟踪 ··43
2.4.1　总体说明 ···43
2.4.2　跟踪 3D 物体 ··45

第 3 章　空间相关内容 ···51
3.1　表面跟踪和运动跟踪 ··51
3.1.1　总体说明 ···51
3.1.2　表面跟踪 ···52

3.1.3　运动跟踪 ··· 55
　3.2　稀疏空间地图 ··· 57
　　　3.2.1　总体说明 ··· 57
　　　3.2.2　建立地图 ··· 60
　　　3.2.3　本地化稀疏空间地图 ··· 66
　3.3　稠密空间地图 ··· 71
　　　3.3.1　总体说明 ··· 71
　　　3.3.2　建立并使用稠密空间地图 ·· 71

第 4 章　屏幕录像 ··· 79

　4.1　总体说明 ·· 79
　　　4.1.1　CameraRecorder 脚本相关 ······································ 79
　　　4.1.2　VideoCameraDevice 游戏对象相关 ·························· 80
　　　4.1.3　VideoRecorder 游戏对象相关 ··································· 80
　　　4.1.4　禁用多线程渲染 ·· 82
　4.2　使用屏幕录像功能 ··· 82

第 5 章　制作涂涂乐和 3D 跟踪物体例子 ·· 89

　5.1　制作涂涂乐 ··· 89
　5.2　场景制作 ·· 91
　5.3　制作 3D 跟踪物体 ··· 95
　　　5.3.1　寻找合适的模型 ·· 95
　　　5.3.2　模型修改 ·· 97
　　　5.3.3　模型导出和转换 ··· 100
　　　5.3.4　纸模转换制作 ··· 103
　　　5.3.5　模型制作 ·· 107
　　　5.3.6　场景制作 ·· 107

第 6 章　稀疏空间地图室内导航原理 ·· 110

　6.1　增强现实室内导航原理说明 ··· 110
　　　6.1.1　基本原理 ·· 110
　　　6.1.2　利用运动跟踪的实现方式 ·· 112
　　　6.1.3　利用稀疏空间地图的实现方式 ································· 113
　6.2　添加虚拟空间场景内容的方式 ·· 114
　6.3　其他 ··· 115

第 7 章　项目准备 ··· 116

　7.1　总体想法 ·· 116
　7.2　难点解决 ·· 117
　　　7.2.1　对象信息保存 ··· 117

		7.2.2 导航实现	122
7.3	项目设计		142
	7.3.1	场景设计	142
	7.3.2	界面设计	144
	7.3.3	开发模式	145
	7.3.4	其他内容	145
7.4	项目搭建		146

第 8 章 调试场景开发 149

8.1	菜单场景开发		149
	8.1.1	场景设置	149
	8.1.2	添加游戏控制脚本	155
	8.1.3	修改设置场景控制脚本	156
8.2	地图场景开发		161
	8.2.1	场景设置	161
	8.2.2	编写代码	164
8.3	模型场景开发		168
	8.3.1	模型移动功能预制件开发	168
	8.3.2	场景设置	182
	8.3.3	添加虚拟物体功能	185
	8.3.4	界面切换和点击选中	189
	8.3.5	删除和保存	192
8.4	关键点场景开发		199
	8.4.1	场景搭建	199
	8.4.2	编写脚本	208
8.5	预备路径场景开发		226
	8.5.1	场景设置	226
	8.5.2	编写脚本	228
8.6	导航场景开发		239
	8.6.1	场景搭建	239
	8.6.2	界面切换和返回	242
	8.6.3	修改显示模型	243
	8.6.4	添加静态模型	247
	8.6.5	添加模型场景对应模型	249
	8.6.6	添加关键点	250
	8.6.7	添加路径	254
	8.6.8	路径导航和显示	256

第 9 章 实际场景开发 264

9.1	菜单场景开发		264

- 9.2 地图场景开发 ··· 266
  - 9.2.1 添加稀疏空间地图游戏对象 ··· 266
  - 9.2.2 修改返回功能 ·· 267
  - 9.2.3 添加保存功能 ·· 271
- 9.3 模型场景开发 ··· 276
  - 9.3.1 场景设置 ·· 276
  - 9.3.2 脚本修改 ·· 276
- 9.4 关键点场景开发 ··· 283
  - 9.4.1 添加平面跟踪图像 ··· 283
  - 9.4.2 脚本准备 ·· 284
  - 9.4.3 脚本修改 ·· 285
- 9.5 路径场景开发 ··· 292
- 9.6 导航场景开发 ··· 293
  - 9.6.1 设置场景 ·· 293
  - 9.6.2 修改导航脚本 ·· 296

第 10 章 调试发布 ·· 305

- 10.1 发布调试应用建立地图 ·· 305
- 10.2 调试错误修改 ·· 306
- 10.3 其他场景设置 ·· 307
- 10.4 最终导航 ·· 308
- 10.5 最终清理 ·· 309

第 1 章 EasyAR 4.0 基础

1.1 EasyAR 4.0 简介

1. 基本介绍

EasyAR 的全称是 EasyAR Sense（SDK），它是视辰信息科技（上海）有限公司的增强现实解决方案系列的子品牌，是国内增强现实 SDK 中使用较多的一款开发包。配套的文档较齐全，同时提供了不错的官方示例，使用起来比较容易上手。EasyAR 4.0 版于 2019 年 12 月发布。

- 官方网址：http://www.easyar.cn/
- 官方 QQ 技术交流群：543115898

EasyAR Sense（SDK）提供了平面图像跟踪（图片识别）、3D 物体跟踪（物体识别）、运动跟踪、稀疏空间地图、稠密空间地图等功能。在官方的示例中，除了识别图片显示模型、播放视频外，还提供了涂涂乐的例子。此外，EasyAR 还提供了手势识别和姿势识别的 SDK。

EasyAR 除了提供 Unity 的开发包，还提供了 Web 的开发包，可以在网页和微信里实现 AR。

2. 版本和功能

EasyAR 4.0 分个人版、专业版和企业版。个人版免费使用，但是有水印（在屏幕右下角）。云识别、手势识别和姿势识别的免费期限是 28 天，空间地图调用限制 100 次/天。其他版本功能收费方式略微复杂，请读者参考官网。

3. 支持平台

支持在以下操作系统中使用：

- Windows 7 及以上版本（7/8/8.1/10）
- Mac OS X
- Android 4.0 及以上版本
- iOS 7.0 及以上版本

Unity 开发支持 Unity 4.6 到 Unity 2018 的版本。

4. 特别说明

EasyAR 4.0 新提供的功能（运动跟踪、稀疏空间地图、稠密空间地图）不能在 Unity 编辑器调试，只能在移动设备上使用，同时对设备有相应的要求。

（1）官方给出了设备支持列表

安卓设备支持列表的网址为 https://help.easyar.cn/EasyAR%20Sense/v4/Guides/EasyAR-Motion-Tracking-Supported-Devices.html。苹果产品中只有支持 ARKit 的设备才支持新功能。在开发相关功能以前，请确定自己的设备支持新功能。

（2）官方演示例子

官方提供了 Unity 所有功能的演示例子。

（3）官方演示例子说明

在页面的下载地址（https://www.easyar.cn/view/download.html#download-nav4）旁边有对每个场景的说明，如图 1-1 所示。

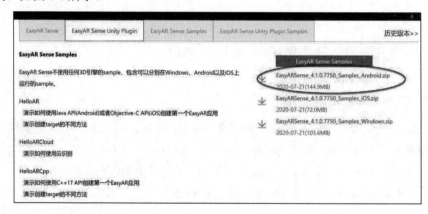

图 1-1

官方演示例子的效果视频为 https://www.bilibili.com/video/BV1Lt4y1U7kf，读者可以查看一下。

1.2 下载导入和基本设置

1.2.1 获取 Key

1. 添加 License Key

License Key 是最基本的 Key，有了这个 Key 才能使用 EasyAR 开发包。打开 EasyAR 网站，单击"开发中心"按钮，如图 1-2 所示。

选中"Sense 授权管理"标签，单击"我需要一个新的 Sense 许可证密钥"按钮，如图 1-3 所示。

在新的界面中，选择"Sense 类型"，添加"应用名称""Bundle ID"和"Package Name"，如图 1-4 所示。其中，"应用名称"是自己取的，"Bundle ID"和"Package Name"相当于一个应用的身份证号，不和其他应用的身份证号冲突即可。

图 1-2

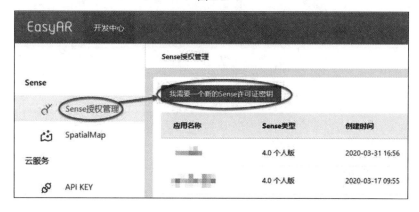

图 1-3

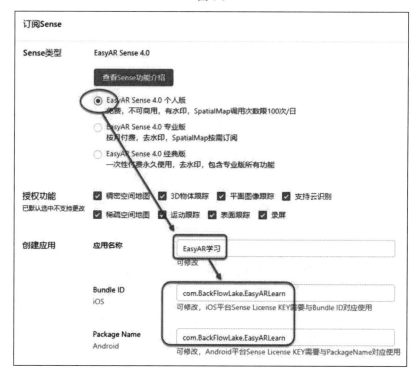

图 1-4

此外还要设置空间地图信息，选择"SpatialMap 区域"，填写"SpatialMap 库名"（自己取即可），然后单击"确认"按钮，如图 1-5 所示。

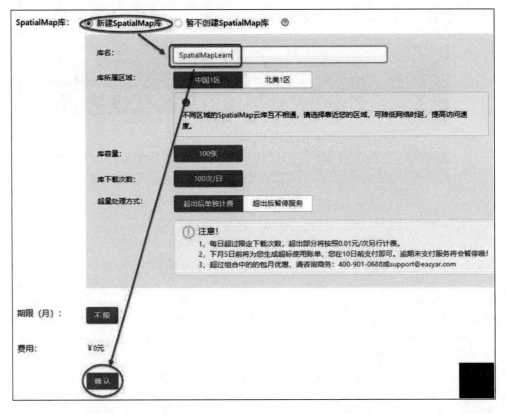

图 1-5

2. 添加云服务 Key

如果需要使用 EasyAR 提供的相关云服务内容，如云识别、稀疏空间地图、手势识别、姿势识别，必须先有这个 Key。

选中 API KEY 标签，单击"创建 API KEY"按钮，如图 1-6 所示。

在新界面中，填写"应用名称"，选择需要的"云服务"内容，填写"应用描述"，然后单击"确认"按钮即可，如图 1-7 所示。

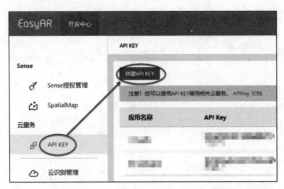

图 1-6

图 1-7

3. 添加云识别 Key

如果想要使用云识别功能，需要添加云识别图库，以获取对应的 Key。选中"云识别管理"标签，单击"新建云识别图库"按钮，如图 1-8 所示。

在新界面中选择"区域"，填写"图库名称"，然后单击"确认"按钮即可，如图 1-9 所示。

图 1-8　　　　　　　　　　　　　　　　　　图 1-9

1.2.2　下载导入

EasyAR 4.0 的下载地址为 https://www.easyar.cn/view/download.html#download-nav2。打开下载页面，找到下载项目，单击即可下载。下载下来的文件是一个 Zip 的压缩包，解压以后就是导入项目用的".unitypackage"文件了，如图 1-10 所示。

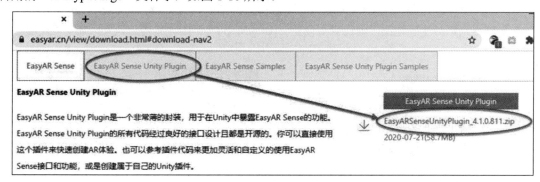

图 1-10

在项目中单击菜单 Assets→Import Package→Custom Package，打开导入窗口，如图 1-11 所示。
在弹出窗口中选中需要导入的文件，单击"打开"按钮，如图 1-12 所示。
在弹出窗口中能看到 EasyAR 导入的内容，单击 Import 按钮即可，如图 1-13 所示。

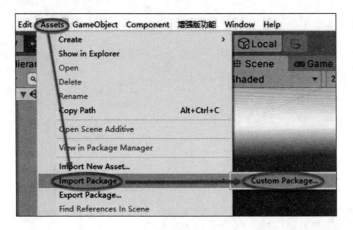

图 1-11

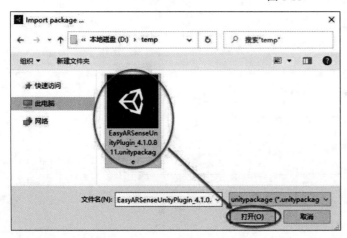

图 1-12

图 1-13

1.2.3 基本设置

使用 EasyAR 首先需要设置 License Key。License Key 和 Bundle ID、Package Name 是绑定的。如果需要使用到稀疏空间地图，则还需要设置 SpatialMap 库；如果需要云识别，还需要设置云识别图库。

1. 设置 License Key

EasyAR 的配置文件是 EasyAR/Resources/EasyAR 目录下的 Settings.asset，可以直接单击文件，也可以单击菜单 EasyAR→Change License Key 打开。

在"开发中心"中，选中"Sense 授权管理"标签，选择其下的项目，如图 1-14 所示。

将 Sense License Key 复制到 EasyAR 配置文件的 EasyAR SDK License Key 属性中，如图 1-15 所示。

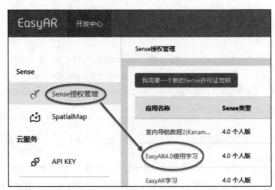

图 1-14

第 1 章　EasyAR 4.0 基础

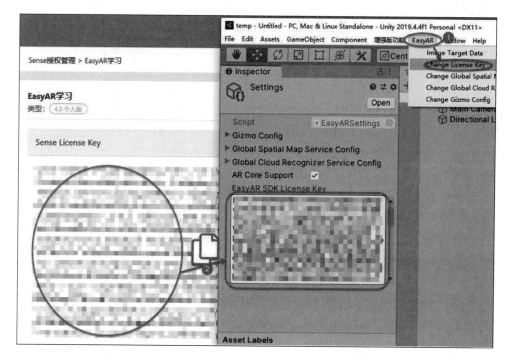

图 1-15

单击 Gizmo Config 还能看到平面图像跟踪和 3D 物体跟踪的设置，默认是全部都选中的状态，一般不需要修改，默认还选中了 AR Core 的支持。如图 1-16 所示。

2. 设置 SpatialMap 库

在"开发中心"中，选中 API KEY 标签，选择其下的项目，如图 1-17 所示。

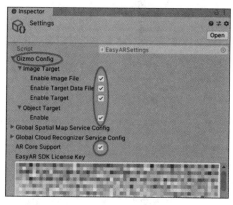

图 1-16

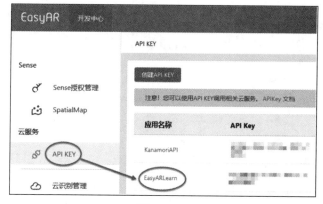

图 1-17

将 API KEY 和 API Secret 复制到 EasyAR 配置文件的 Global Spatial Map Service Config 项目下对应的属性中，如图 1-18 所示。

在"开发中心"中，选中 SpatialMap 标签，选择其下的项目，如图 1-19 所示。

选中"密钥"标签，将下面的 SpatialMap AppId 的内容复制到 Sparse Spatial Map App ID 属性中，如图 1-20 所示。

7

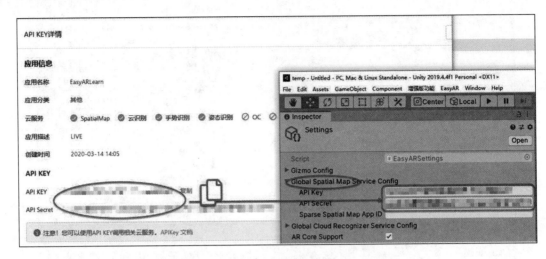

图 1-18

图 1-19

图 1-20

3. 设置云识别库

在"开发中心"中,选中 API KEY 标签,选择其下的项目,如图 1-21 所示。

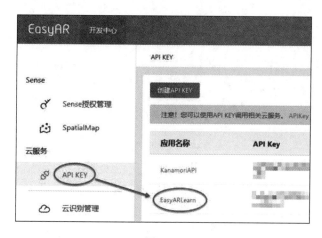

图 1-21

将 API KEY 和 API Secret 复制到 EasyAR 配置文件的 Global Cloud Recognizer Service Config 项目下对应的属性中，如图 1-22 所示。

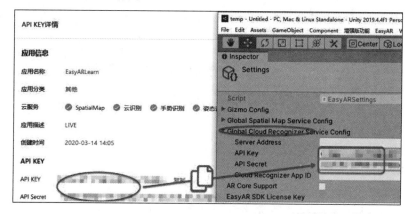

图 1-22

在"开发中心"中，选中"云识别管理"标签，选择其下的项目，如图 1-23 所示。

图 1-23

选中"密钥"标签，复制 CRS AppId 的内容到 Cloud Recognizer App ID 属性中为其赋值，如图 1-24 所示。

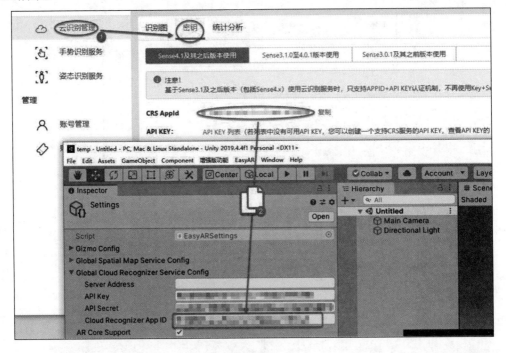

图 1-24

复制 Cloud URLs 下的 Client-end (Target Recognition) URL 内容到 Server Address 属性中，如图 1-25 所示。

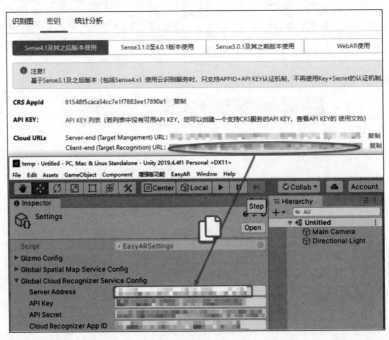

图 1-25

4．其他设置

单击菜单 File→Build Settings，打开 Build Settings 窗口。选中 Platform 下的 Android 标签，将发布平台设置为安卓。单击 Player Settings 打开发布相关的设置。在 Inspector 窗口中，选中 Other Settings...标签，将 Package Name 复制到对应属性中。

这里的 Package Name 和 License Key 必须是对应的，即在同一个授权下，如图 1-26 所示。

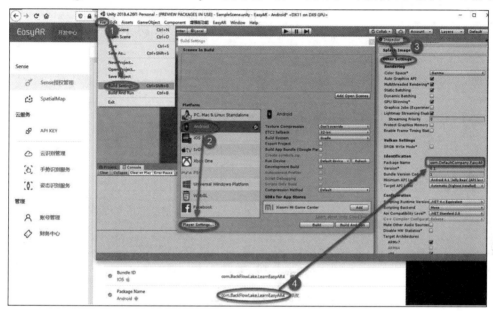

图 1-26

安卓版本至少需要 API Level 17，如果用到运动跟踪，则至少需要 API Level 24，如图 1-27 所示。

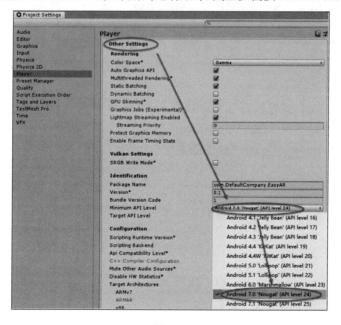

图 1-27

场景中至少需要一个标签为 MainCamera 的摄像机，才能正确显示摄像头拍摄到的内容。通常情况下，将默认的 Main Camera 游戏对象的 Clear Flags 值设置为 Solid Color 即可，如图 1-28 所示。

如果运行后能跟踪功能正常，但是背景不是摄像头拍摄到的内容，通常原因是此处没有设置。

关于 Unity 版本，建议初学者一定使用 LTS 长期支持版。这种版本的 Bug 最少。Unity 并不像其他软件，Unity 最新版更多的是用来体验其最新功能，开发和学习都请选用 LTS 版。本书所使用的 Unity 版本是 2018.4，如图 1-29 所示。

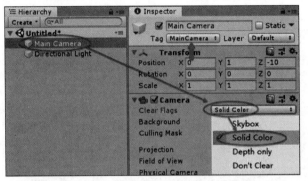
图 1-28

图 1-29

1.3　EasyAR 的基本结构

EasyAR 的基本结构包括一个 EasyAR 游戏对象、一个 RenderCamera 游戏对象和一个 CameraDevice 游戏对象。其中，CameraDevice 可以是 VideoCameraDevice，也可以是 VIOCameraDevice。

CameraDevice 游戏对象负责控制摄像头，RenderCamera 游戏对象负责将摄像头拍摄到的内容显示出来。

VIOCameraDevice 功能更强，但是对设备有要求。VideoCameraDevice 兼容性更强。如果使用到运动跟踪，则一定要使用 VIOCameraDevice；其他情况则可以使用 VideoCameraDevice 以提高对设备的兼容。

EasyAR 的基本结构如图 1-30 所示。

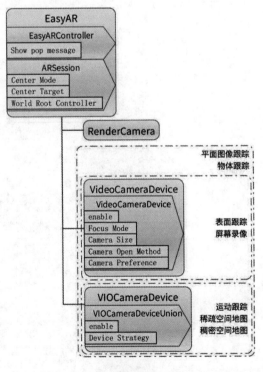
图 1-30

1.3.1 EasyAR 游戏对象下的设置

（1）Show pop message 选项

EasyARController 的 Show pop message 选项的作用是是否在平面上显示 EasyAR 的提示信息。当选中的时候，如果 EasyAR 有错误信息就会直接显示在屏幕上，通常不需要修改，如图 1-31 所示。

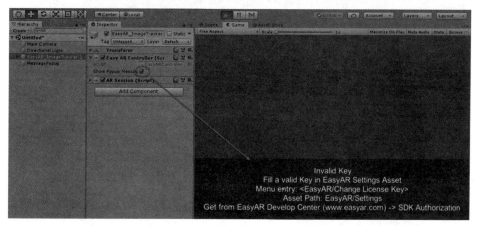

图 1-31

（2）Center Mode 选项

ARSession 的 Center Mode 选项用来控制在场景中哪个游戏对象是参照对象，如图 1-32 所示。

选中 Camera，在进行平面跟踪和物体跟踪的时候，Main Camera（也就是场景中的摄像机游戏对象）的位置和角度固定不变，而被跟踪对象的位置和角度依据设备的移动情况进行变化。

选中 First Target 或者 Specific Target，在进行平面跟踪和物体跟踪的时候，第一个被识别的游戏对象或者指定的游戏对象位置和角度固定不变，而其他的识别对象和摄像机游戏对象的位置和角度依据设备的移动情况进行变化。

图 1-32

选中 World Root，需要设置 World Root Controller 属性，在跟踪开始的时候，World Root Controller 属性指定的游戏对象位置和角度固定不变，而其他的游戏对象的位置和角度依据设备的移动情况进行变化。

1.3.2 VideoCameraDevice 游戏对象下的设置

（1）enable 属性

enable 属性用于打开或关闭摄像头，通常在程序控制中使用，如图 1-33 所示。

（2）Focus Mode 选项

该选项用于设置摄像头的对焦模式，默认值是 Continousauto，如图 1-34 所示，具体选项说明如表 1-1 所示。

图 1-33

图 1-34

表 1-1 摄像头的对焦模式

名 称	值	说 明
Normal	0	常规对焦模式,在这个模式下需要调用 autoFocus 来触发对焦
Continousauto	2	连续自动对焦模式
Infinity	3	无穷远对焦模式
Macro	4	微距对焦模式。在这个模式下需要调用 autoFocus 来触发对焦
Medium	5	中等距离对焦模式

(3) Camera Open Method 选项

该选项用于设置使用哪个摄像头来获取环境图像,可以通过类型或者序号来设置,如图 1-35、图 1-36 所示。

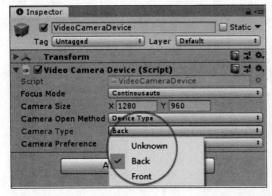

图 1-35

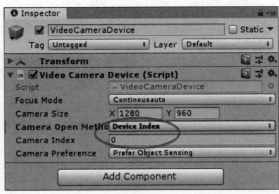

图 1-36

（4）Camera Preference 选项

该选项默认选择 Prefer Object Sensing，只有当使用到表面跟踪的时候才选择 Prefer Surface Tracking，如图 1-37 所示。

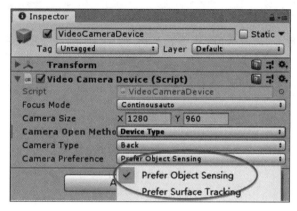

图 1-37

1.3.3　VIOCameraDevice 游戏对象下的设置

Device Strategy 选项是用于设置 VIO（Visual-Inertial Odometry）视觉惯性里程计的使用。System VIO 指的是设备中安装的 ARCore 或者 ARKit，如图 1-38 所示。

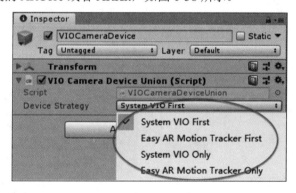

图 1-38

VIOCameraDevice 游戏对象的 enable 属性同样可以用来打开和关闭摄像头。

第 2 章
图像和物体跟踪

2.1 平面图像跟踪

2.1.1 总体说明

平面图像跟踪对图像有一定的要求,纹理细节丰富,并且纹理不是简单重复,长宽比不能太大。

官方提供了图像检测工具(https://www.easyar.cn/targetcode.html),将图像上传到指定地址,可以知道可识别度,如图 2-1 所示。其中,1 星、2 星是不能识别,3 星是较难识别,4 星、5 星是易识别。

平面图像跟踪主要是 ImageTracker 和 ImageTarget 这两个游戏对象,每个 ImageTarget 对应一个被跟踪的图像,场景中也可以同时出现多个 ImageTracker。平面图像跟踪结构如图 2-2 所示。

图 2-1

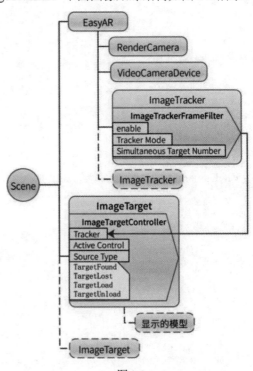

图 2-2

1. ImageTracker 游戏对象相关

ImageTracker 游戏对象的主要属性如图 2-3 所示。

图 2-3

（1）enable 属性

enable 属性可以用于获取当前跟踪器状态，也可以用于启用/禁用当前跟踪器，从而实现启用/禁用平面图像识别的功能。

（2）Tracker Mode 选项

该选项设置跟踪的时候是质量优先还是性能优先，默认是质量优先（Prefer Quality）。

（3）Simultaneous Target Number 设置

该设置指定当前的跟踪器同时跟踪目标的最大数量。一个场景中，能同时被跟踪的图像数量是所有 ImageTracker 的 Simultaneous Target Number 属性值的和。

2. ImageTarget 游戏对象相关

ImageTarget 游戏对象的主要属性如图 2-4 所示。

图 2-4

（1）Tracker 设置

每个 ImageTarget 游戏对象必须指定一个 ImageTracker 游戏对象才能被跟踪，可以通过修改该设置实现对图像的加载和卸载。

可以通过设置该属性为 null 实现卸载对应目标对象，设置该属性为具体的 ImageTracker 游戏对象实现加载对应目标对象。

（2）Active Control 选项

该选项用于设置 ImageTarget 游戏对象是否激活。

当选择 Hide When Not Tracking 的时候，只有图像被跟踪时，ImageTarget 游戏对象才被激活；图像没有被跟踪，则 ImageTarget 游戏对象不被激活。默认为该选项。

当选择 Hide Before First Found 的时候，当图像第一次被跟踪以后，ImageTarget 游戏对象就被激活，之后一直处于激活状态。

当选择 None 的时候，ImageTarget 游戏对象始终被激活。

当一个 ImageTarget 游戏对象被激活但是又没有被跟踪的时候，其位置和角度不会变化。

（3）Source Type 选项

该选项用于设置跟踪类型，除了可以直接跟踪图片 Image File，还可以跟踪只包含关键信息文件大小小很多的 Target Data File。

Target 用来获取云识别结果。

（4）事件

ImageTargetController 类提供了 4 个事件，分别是图像被识别 TargetFound、被识别图像从视野消失 TargetLost、图像加载完成 TargetLoad 和图像卸载完成 TargetUnload。

通过订阅这些事件就能实现对应的程序控制。

```
controller.TargetFound += () =>
{
   ...
};
controller.TargetLost += () =>
{
   ...
};
controller.TargetLoad += (Target target, bool status) =>
{
   ...
};
controller.TargetUnload += (Target target, bool status) =>
{
   ...
};
```

2.1.2 跟踪单个图像

1. 项目准备

（1）新建一个空的 Unity 项目，删除项目中的默认内容。

（2）导入 EasyAR SDK，设置 License Key。

（3）设置导出平台为安卓平台，并设置 Package Name 以及导出的安卓版本。

（4）添加目录 StreamingAssets，用于放置本地的识别对象信息。

（5）再添加一个项目目录，这里是 EasyAR 4Learn，用于放置该项目的所有内容。

最终操作结果如图 2-5 所示。

2. 添加基本内容

在 EasyAR 4Learn/Scenes 目录下新建一个场景，保存为 ImageTarget-Base。设置场景中 Main Camera 的 Clear Flags 属性为 Solid Color，如图 2-6 所示。

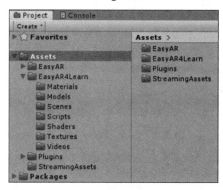
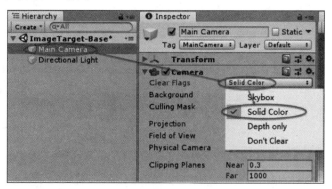

图 2-5　　　　　　　　　　　　　　　图 2-6

将 EasyAR/Prefabs/Composites 目录下的"EasyAR_ImageTracker-1"预制件拖到场景中，如图 2-7 所示。

将 EasyAR/Prefabs/Primitives 目录下的 ImageTarget 预制件拖到场景中，如图 2-8 所示。

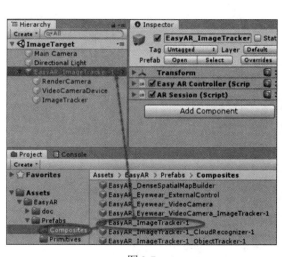

图 2-7　　　　　　　　　　　　　　　图 2-8

3. 设置利用图片文件进行跟踪

将要跟踪的图片 bus.jpg 拖动到 StreamingAssets 目录下。

设置 ImageTarget 游戏对象的 Source Type 属性为 Image File，通过图像进行跟踪。

设置 Path Type 为 Streaming Assets，设置 Path 为 bus.jpg，即跟踪图像相对路径，设置 Name 为 bus，设置 Scale 属性为 0.2，如图 2-9 所示。

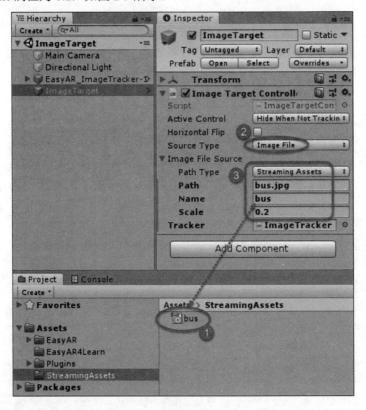

图 2-9

注意，这里的 Scale 大小是指图像在被跟踪的时候在现实空间的宽度，单位为米。设置为 0.2 即在现实空间中该图像大小为 0.2 米，差不多是一把小学生用的尺子的长度。

在 ImageTarget 游戏对象下添加一个方块，如图 2-10 所示。

图 2-10

运行时，当视野中有图像出现的时候就会在对应位置显示一个方块，如图 2-11 所示。

图 2-11

4. 设置利用 Data 文件进行跟踪

单击菜单 EasyAR→Image Target Data，如图 2-12 所示。

在弹出窗口中，设置 Generate From 为 Image。将要跟踪的图像拖到 Image Path 中。设置 Name 和 Scale 属性，然后单击"Generate"按钮，如图 2-13 所示。之后，默认会在项目的 SteamingAssets 目录下生成.etd 文件。

这里的 Scale 和前面的一样，是指图像在被跟踪的时候在现实空间的宽度，单位为米。

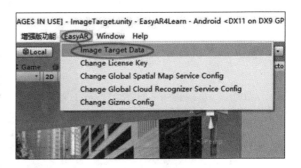

图 2-12

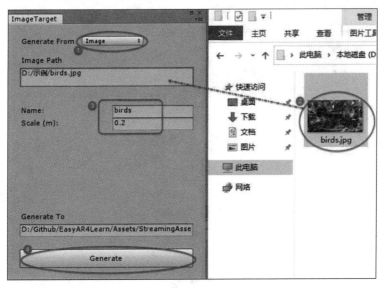

图 2-13

将 EasyAR/Prefabs/Primitives 目录下的 ImageTarget 预制件拖到场景中。

设置 ImageTarget(1)游戏对象的 Source Type 属性为 Target Data File，通过数据文件进行跟踪；设置 Path Type 为 Streaming Assets，设置 Path 为 birds.etd，即跟踪数据文件相对路径；在 ImageTarget(1)游戏对象下添加一个球体，如图 2-14 所示。

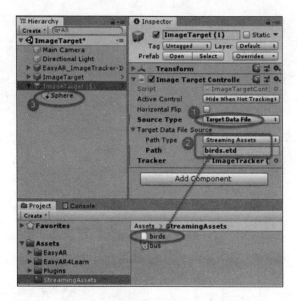

图 2-14

运行时,当视野中有图像出现的时候,就会在对应位置显示一个球体,如图 2-15 所示。

图 2-15

2.1.3 跟踪多个图像

1. 单个 Tracker

将 ImageTarget-Base 场景另存为 ImageTarget-MultiTarget。打开 ImageTarget-MultiTarget 场景,修改 ImageTracker 游戏对象的 Simultaneous Target Number 值,将其从默认 1 改为 2,如图 2-16 所示。

图 2-16

运行时可以同时跟踪 2 个图像，如图 2-17 所示。

图 2-17

这是最常用的方法，即用一个 Tracker 同时跟踪多个图像。当视野中的图像数量小于等于跟踪数量的时候，所有图像都会被跟踪。当视野中的图像数量大于跟踪数量的时候，最先被跟踪的图像会被跟踪。或者说，哪些图像会被跟踪，虽可影响但不可控。

2. 多个 Tracker

将 ImageTarget-MultiTarget 场景另存为 ImageTarget-MultiTarget MultiTracker。打开 ImageTarget-MultiTargetMultiTracker 场景，重新命名原有的 Tracker 和 Target。

将原有的 ImageTracker 名称改为 ImageTracker-A。原有的 Target 后面也添加"-A"标识，即原有的两个 Target 是由 A Tracker 进行跟踪，如图 2-18 所示。

图 2-18

在场景中再添加一个 Tracker，设置 Simultaneous Target Number 值为 2。在其名称后面添加"-B"，即标识为 B Tracker，如图 2-19 所示。

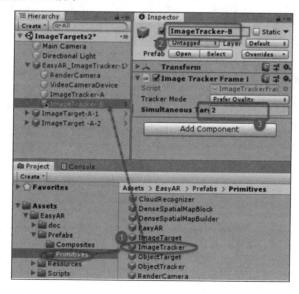

图 2-19

再添加 4 个 ImageTarget，设置为跟踪不同的图像，Tracker 属性都设置为 B Tracker，即后添加的 Tracker，如图 2-20 所示。

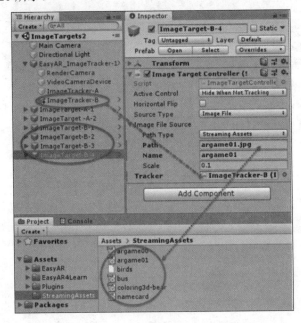

图 2-20

运行场景，视野中的 6 个图像能同时被跟踪 4 个，即每个 Tracker 跟踪 2 个图像，如图 2-21 所示。设置为第一个 Tracker（A）的 2 个图像始终都能被跟踪，而设置为第二个 Tracker（B）的 4 个图像只有 2 个能被跟踪，具体是哪 2 个不可控。

图 2-21

3. 跟踪同一图像的不同副本

将 ImageTarget-MultiTargetMultiTracker 场景另存为 ImageTarget-MultiSameTarget。打开 ImageTarget-MultiSameTarget 场景，将所有 ImageTarget 的 Path 属性都设置为 namecard.jpg，即都是跟踪同一个图像，如图 2-22 所示。其中，2 个 Target 使用 A Tracker，4 个 Target 使用 B Tracker。设置 A Tracker 的 Simultaneous Target Number 值为 1、B Tracker 的 Simultaneous Target Number 值为 4。在 ImageTarget 下设置不同的模型以示区分。

第 2 章　图像和物体跟踪

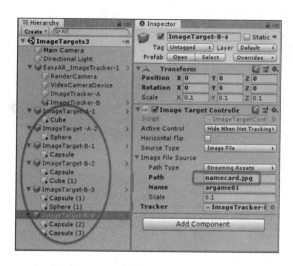

图 2-22

运行场景，当视野中有 9 个图像的时候，随机跟踪其中的 5 个图像。使用 B Tracker 的所有 Target 都被跟踪了，使用 A Tracker 的 Target 随机跟踪一个，如图 2-23 所示。

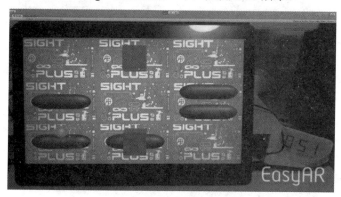

图 2-23

2.1.4　平面图像跟踪程序控制

这里，通过在场景中添加 2 个切换（Toggle）来控制 ImageTracker 和 ImageTarget，同时在触发事件的时候将内容打印到控制台。

1. 添加切换

将 ImageTarget-MultiTargetMultiTracker 场景另存为 ImageTarget-Program，打开 ImageTarget-Program 场景。

单击菜单 GameObject→UI→Toggle，添加切换，如图 2-24 所示。

添加 2 个切换，修改名称，设置位置和大小，这里设置在屏幕左下方，如图 2-25 所示。

2. 添加脚本

在场景中添加一个空的游戏对象并命名为 SceneMaster。在目录 EasyAR 4Learn/Scripts 中添加一个脚本 ImageTargetsController，并将其拖到 SceneMaster 游戏对象下成为其组件，如图 2-26 所示。

25

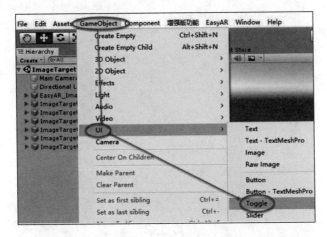

图 2-24

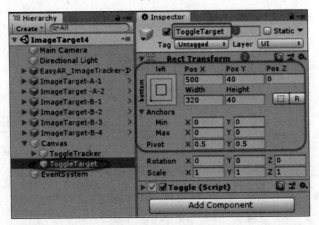

图 2-25

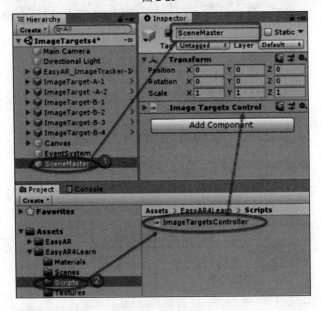

图 2-26

脚本内容如下：

```csharp
public ImageTrackerFrameFilter tracker;
public ImageTargetController targetController;
void Awake()
{
    if (targetController)
    {
        targetController.TargetFound += () =>
        {
            Debug.LogFormat(...);//当图像被跟踪到
        };
        targetController.TargetLost += () =>
        {
            Debug.LogFormat(...);//当图像从视野消失
        };
        targetController.TargetLoad += (Target target, bool status) =>
        {
            Debug.LogFormat(...);//加载图像
        };
        targetController.TargetUnload += (Target target, bool status) =>
        {
            Debug.LogFormat(...);//卸载图像
        };
    }
}
public void SetTracker(bool status)
{
    tracker.enabled = status;//设置Tracker
}
public void SetTarget(bool status)
{
    if (status)
    {
        targetController.Tracker = tracker;//加载图像
    }
    else
    {
        targetController.Tracker = null;//卸载图像
    }
}
```

3. 设置脚本

将 ImageTracker 和对应的 ImageTarget 赋值到脚本对应的属性中。这里是将 ImageTracker-A 游戏对象拖到脚本的 Tracker 属性中为其赋值，将 ImageTarget-A-1 游戏对象拖到脚本 Target Controller 中为其赋值，如图 2-27 所示。

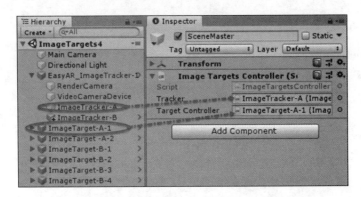

图 2-27

选中 ToggleTracker 游戏对象，单击 On Value Changed 标签下的 "+" 按钮，添加 On Value Changed 事件；将 SceneMaster 游戏对象拖到 On Value Changed 下的框中，设置事件响应的方法来自 SceneMaster 游戏对象；单击右边的下拉菜单，选择 ImageTargetsController 下的 SetTracker，即设置事件响应的方法是 ImageTargetsController 脚本下的 SetTracker 方法，如图 2-28 所示。

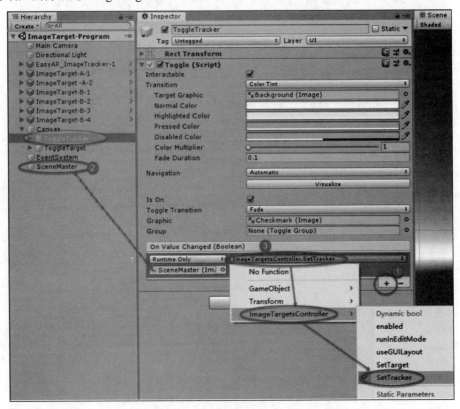

图 2-28

用同样的方法设置 ToggleTarget 游戏对象的 On Value Changed 事件响应的方法，把它设置为 SceneMaster 游戏对象下 ImageTargetsController 组件的 SetTarget 方法。

运行场景，当取消 Tracker 切换的时候，对应的 2 个图像都不被跟踪，如果只取消 Target 切换，只有对应的（bus.jpg）图像不被跟踪。在控制台能看到打印的事件信息，如图 2-29 所示。

图 2-29

2.2 平面图像跟踪扩展内容

2.2.1 视频播放

平面图像识别以后,播放视频也是一种经常被用到的增强现实的表现方式。通常是截取视频第一帧的图片作为识别图片,识别以后再播放视频。这样就给人一种图片动起来的错觉。

1. 添加并设置基础内容

在目录 EasyAR 4Learn/Scenes 下新建场景并命名为 ImageTarget-Video;在场景中设置 Main Camera 的 Clear Flags 属性为 Solid Color;将 EasyAR/Prefabs/Composites 目录下的 EasyAR_ImageTracker-1 预制件拖到场景中;将 EasyAR/Prefabs/Primitives 目录下的 ImageTarget 预制件拖到场景中,一共拖 2 个;设置 ImageTarget 的识别图片,如图 2-30 所示。

2. 添加并设置视频播放的游戏对象

选中一个 ImageTarget,单击鼠标右键,在弹出的菜单中选择 3D Object→Plane,为 ImageTarget 对象添加一个平面作为其子游戏对象,如图 2-31 所示。

设置 Plane 游戏对象的角度为 "90, 0, 180",并修改其 Scale 属性,将其大小设置得和视频图片大小一样,如图 2-32 所示。

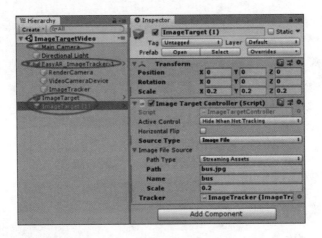

图 2-30

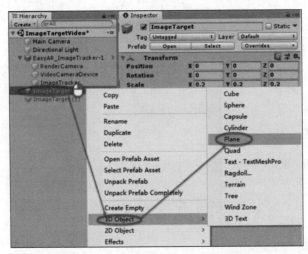

图 2-31

图 2-32

3. 添加并设置视频播放组件

选中 Plane 游戏对象，单击菜单 Component→Video→Video Player，为其添加一个视频播放组件，如图 2-33 所示。

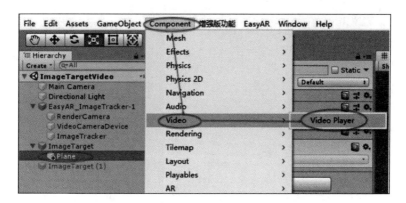

图 2-33

将一段视频拖到目录 EasyAR 4Learn/Videos 下，导入为资源，并将其拖动到 Video Player 组件的 Video Clip 属性中，如图 2-34 所示。

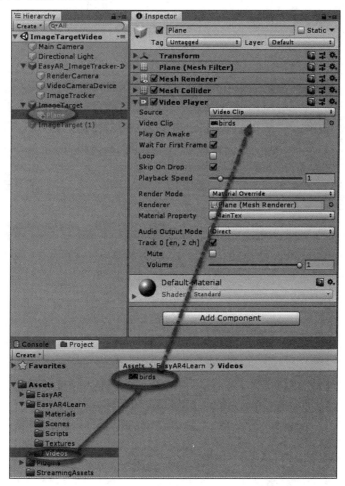

图 2-34

运行，当识别到图片的时候就会在其上播放对应的视频，如图 2-35 所示。其中，视频画面的长宽由 Plane 游戏对象决定。

图 2-35

视频不仅可以在平面上播放,也可以在其他形状的 3D 物体上播放。

在另外一个 ImageTarget 对象下添加一个方块作为视频播放的游戏对象,设置方块的大小和角度,添加 Video Player 组件并设置组件的 Video Clip 属性,如图 2-36 所示。

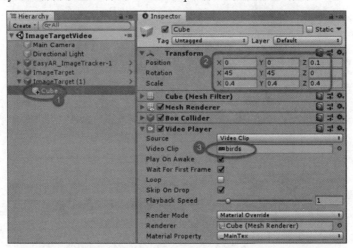

图 2-36

运行,当识别出图片以后就可以看到一个方块,在方块的表面播放视频,如图 2-37 所示。

图 2-37

2.2.2 涂涂乐

涂涂乐也是平面图像识别常用的一种表现形式,通过自己涂色来决定模型显示的表面颜色,有很强的互动功能,在一些展会和幼教产品中经常能见到。

1. 导入相关内容

将 bear.fbx 文件拖到 EasyAR 4Learn/Models 目录下,从官方示例中导入小熊的模型,如图 2-38 所示。

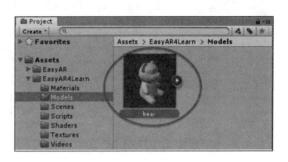

图 2-38

将 Coloring3D.shader 文件拖到 EasyAR 4Learn/Shaders 目录下,从官方示例中导入 Coloring3D 着色器,如图 2-39 所示。

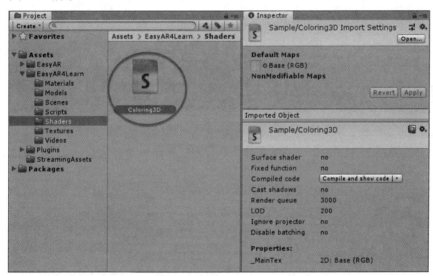

图 2-39

将 Sample_TextureSample.mat 文件拖到 EasyAR 4Learn/Materials 目录下,从官方示例中导入 Sample_Texturesample 材质;同时设置材质对应的着色器是 Sample/Coloring3D,如图 2-40 所示。

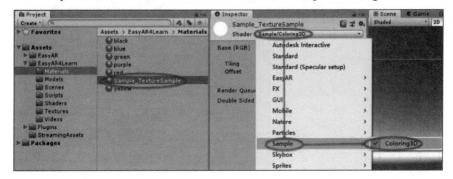

图 2-40

将 Coloring3D.cs 文件拖到 EasyAR 4Learn/Scripts 目录下，导入脚本。这个脚本是用官方示例脚本改写的，区别只是去掉了对按钮的绑定，如图 2-41 所示。

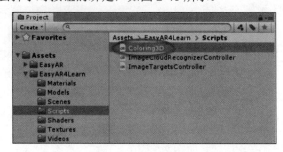

图 2-41

2. 添加并设置基础内容

在 EasyAR 4Learn/Sences 目录下，新建场景并命名为 ImageTarget-Coloring3D。在场景中设置 Main Camera 的 Clear Flags 属性为 Solid Color。

将EasyAR/Prefabs/Composites目录下的EasyAR_ImageTracker-1预制件拖到场景中；将EasyAR/Prefabs/Primitives目录下的ImageTarget预制件拖到场景中；导入识别图片并设置ImageTarget的识别图片为coloring3d-bear.jpg，这里的识别图片必须和3D模型对应，是一个特殊UV展开后的纹理贴图，如图2-42所示。

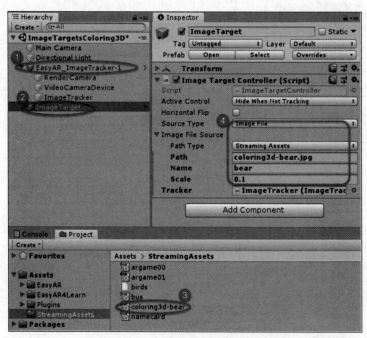

图 2-42

3. 添加并设置模型

在 ImageTarget 游戏对象下添加模型 bear，设置模型的方向和位置，修改模型的材质为 Sample_TextureSample，如图 2-43 所示。

第 2 章 图像和物体跟踪

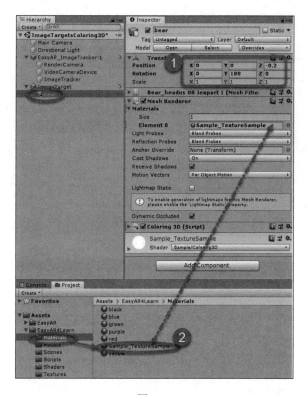

图 2-43

将 Coloring3D 脚本拖到模型上成为其组件，将 RenderCamera 游戏对象拖到脚本组件的 Camera Renderer 属性中，如图 2-44 所示。

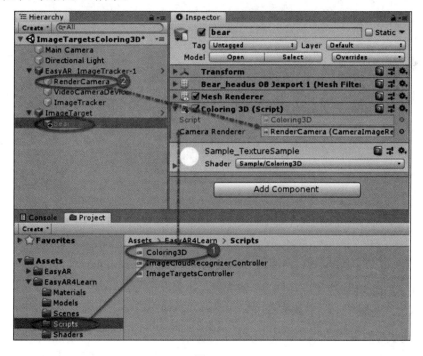

图 2-44

35

此时运行，修改识别图片中小熊的颜色，对应模型的颜色也会随之修改，如图 2-45 所示。

图 2-45

4. 设置锁定功能

单击菜单 GameObject→UI→Button，在场景中添加按钮并命名为 ButtonFreeze；为按钮添加 On Click()事件响应；将小熊模型对象（bear 游戏对象）拖到 On Click()事件响应中，设置事件响应的方法是 Coloring3D 脚本下的 Freeze()方法，如图 2-46 所示。

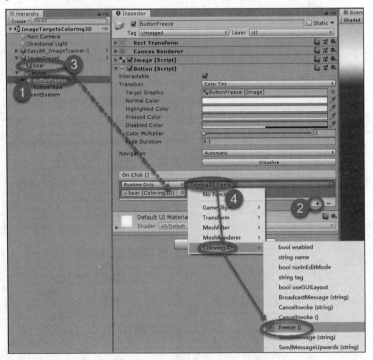

图 2-46

用相同的方法在场景中添加按钮并命名为 ButtonThaw；为按钮添加"On Click()"事件响应；将小熊模型对象拖到 On Click()事件响应中，设置事件响应的方法是 Coloring3D 脚本下的 Thaw()方法。

此时运行，修改了小熊颜色以后，在对应模型颜色也修改了之后，单击 Freeze 按钮。再次识别没涂色的图片，模型依然能显示修改以后的效果，如图 2-47 所示。

锁定效果在重启或者单击 Thaw 按钮后取消。

图 2-47

2.3 云 识 别

2.3.1 总体说明

EasyAR 云识别的图库默认 10 万张图片，同时提供了对图库图片进行操作的 API 接口。这里主要说明的是如何实现云识别。云识别的基本结构如图 2-48 所示。

云识别主要是在平面图像识别的 Tracker 预制件中添加了 Cloudrecognizer 游戏对象。通过设置该游戏对象的 enable 属性，可以实现云识别功能的启用和禁用；通过 UseGlobalServiceConfig 属性可以单独配置云识别的图库；通过订阅 CloudUpdate 事件可以获取识别状态和被识别到的平面图像的 Target。

云识别同样需要用到 ImageTarget 游戏对象，只不过它通常是动态生成的。

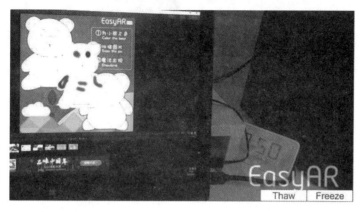

图 2-48

2.3.2 上传图片

云识别前需要将图片上传到图库。

进入 EasyAR 开发中心，选中"云识别管理"标签，单击已经创建好的一个图库，如图 2-49 所示。

进入图库后，能看到已经上传到图库的图片。单击"识别图"标签下的"上传识别图"按钮，如图 2-50 所示。

在弹出窗口中填写"识别图片名称"；单击"浏览"按钮选中要上传的图片；设置宽度，这里宽度的单位是厘米，和 Unity 中的单位不一样；设置完成以后单击"确认"按钮，如图 2-51 所示。

图 2-49

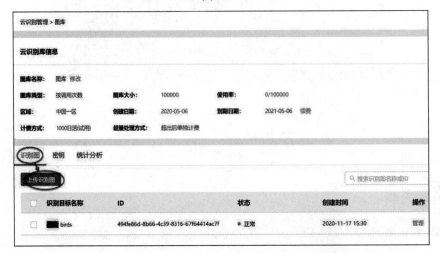

图 2-50

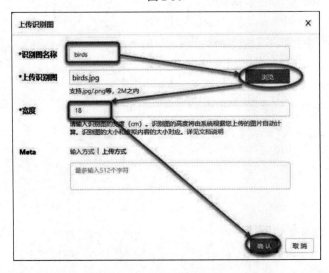

图 2-51

在图库界面中,单击上传到图库的图片可以看到图片的具体信息。其中除了图片的名称等基本信息,还包括图片的"可识别度"和"可跟踪度",可以用于了解图像是否容易被识别和跟踪,如图 2-52 所示。

图 2-52

2.3.3 添加基本内容

云识别基本内容很少，主要内容集中在代码里。

在 EasyAR 4Learn/Scenes 目录下新建一个场景并命名为 ImageTarget-Cloud。设置 Main Camera 游戏对象清除标志 Clear Flags 为 Solid Color。将 EasyAR/Prefabs/Composites 目录下的 EasyAR_ImageTracker-1_CloudRecognizer-1 预制件拖到场景中，如图 2-53 所示。

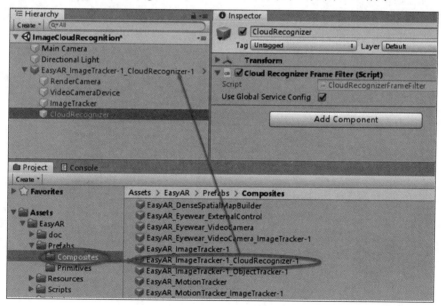

图 2-53

单击菜单 GameObject→UI→Text 和 GameObject→UI→Toggle，在场景中添加一个文本显示 Text 和一个切换按钮 Toggle，新增一个脚本并将其拖到一个空的游戏对象中，如图 2-54 所示。

设置切换按钮 Toggle 的 On Value Changed 响应对象是 CloudRecognizer 游戏对象的 CloudRecognizerFrameFilter 脚本组件的 enabled 属性，如图 2-55 所示。

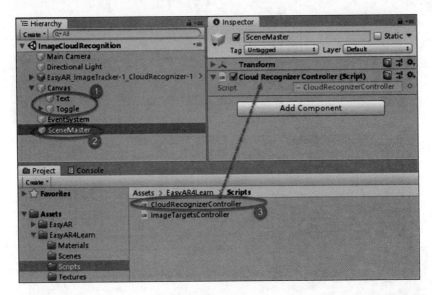

图 2-54

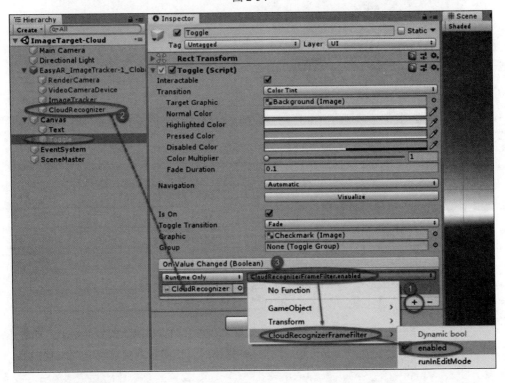

图 2-55

2.3.4 相关程序控制

修改脚本，关键代码如下：

```
void Awake()
{
    ...
```

```
    cloudRecognizer = FindObjectOfType<CloudRecognizerFrameFilter>();

    cloudRecognizer.CloudUpdate += (status, targets) =>
    {
        text.text = "Cloud Recognizer status " ...;
        foreach (var t in targets)
        {
            text.text = text.text +...;
        }
    };
}
```

在 Awake 事件中，订阅 CloudRecognizerFrameFilter 脚本的 CloudUpdate 事件。该事件每秒运行 2 次左右，每次都会返回状态和被识别的目标队列。

运行时，当识别到图像就能显示该图像在图库的名称和 UID，如图 2-56 所示。

图 2-56

每次 CloudUpdate 事件就是一次云识别的计数，而不是识别到一张图片进行一次计数，所以每次云识别可能都会有数个甚至数十个计数。因此，减少无意义的计数很重要。

在官方的示例中，启动以后，在 Awake 事件中会读取作为本地缓存的.etd 文件来加载识别目标图像。

```
if (UseOfflineCache)
{
    if (string.IsNullOrEmpty(OfflineCachePath))
    {
        OfflineCachePath = Application.persistentDataPath + "/CloudRecognizerSample";
    }
    if (!Directory.Exists(OfflineCachePath))
    {
        Directory.CreateDirectory(OfflineCachePath);
    }
    if (Directory.Exists(OfflineCachePath))
```

```csharp
    {
        var targetFiles = Directory.GetFiles(OfflineCachePath);
        foreach (var file in targetFiles)
        {
            if (Path.GetExtension(file) == ".etd")
            {
                LoadOfflineTarget(file);
            }
        }
    }
}
```

在 CloudUpdate 订阅事件中，遍历读取到的目标图像。

```csharp
foreach (var target in targets)
{
    var uid = target.uid();
    if (loadedCloudTargetUids.Contains(uid))
    {
        continue;
    }
    LoadCloudTarget(target.Clone() as ImageTarget);
}
```

如果目标图像没有被缓存，则用 Target 类的 Save 方法将目标图像保存成本地的.etd 文件。

```csharp
private void LoadCloudTarget(ImageTarget target)
{
    ...
    if (UseOfflineCache && Directory.Exists(OfflineCachePath))
    {
        if (target.save(OfflineCachePath + "/" + target.uid() + ".etd"))
        {
            cachedTargetCount++;
        }
    }
}
```

在设置跟踪图像时订阅 TargetFound 方法，只要有一个图像被跟踪就禁用 CloudRecognizer 组件，达到停止云识别的目的。

```csharp
private void LoadTargetIntoTracker(ImageTargetController controller)
{
    controller.Tracker = tracker;
    controller.TargetFound += () =>
    {
        cloudRecognizer.enabled = false;
    };
```

```
controller.TargetLost += () =>
{
    cloudRecognizer.enabled = true;
};
}
```

官方云识别示例结构如图 2-57 所示。

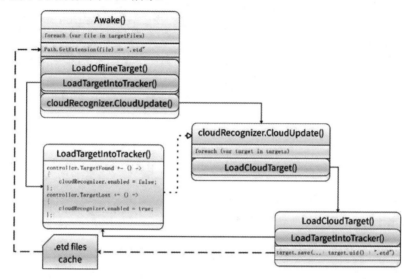

图 2-57

2.4　3D 物体跟踪

2.4.1　总体说明

3D 物体跟踪总体上和平面图像跟踪差不多，包括程序控制、识别多个对象，区别只是目标对象不同。

3D 物体跟踪对 3D 物体的纹理（也就是表面图案的丰富程度）是有要求的。纹理如果是简单的色块组成的，效果并不很好。

官方对被跟踪的 3D 物体的详细要求可以查看官方网站的文档（https://help.easyar.cn/EasyAR%20Sense/v4/Guides/EasyAR-3D-Object-Tracking.html）。

3D 物体跟踪主要是 ObjectTracker 和 ObjectTarget 这两个游戏对象，每个 ObjectTarget 对应一个被跟踪的 3D 物体，场景中也可以同时出现多个 ObjectTracker。3D 物体跟踪结构如图 2-58 所示。

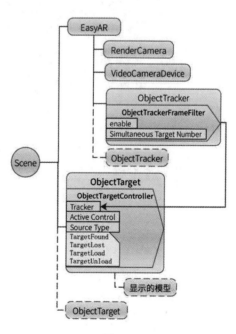

图 2-58

1. ObjectTracker 游戏对象相关

ObjectTracker 游戏对象的主要属性如图 2-59 所示。

图 2-59

（1）enable 属性

enable 属性可以用于获取当前跟踪器状态，也可以用于启用/禁用当前跟踪器。

（2）Simultaneous Target Number 设置

该设置指定当前的跟踪器同时跟踪目标的最大数量。

2. ObejctTarget 游戏对象相关

ObjectTarget 游戏对象的主要属性如图 2-60 所示。

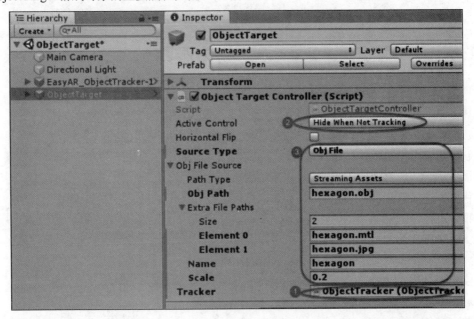

图 2-60

（1）Tracker 设置

每个 ObjectTarget 游戏对象必须指定一个 ObjectTracker 游戏对象才能被跟踪，可以通过修改该设置实现对 3D 物体的加载和卸载。

可以通过设置该属性为 null 实现卸载对应目标对象，设置该属性为具体的 ObjectTracker 游戏对象以实现加载对应目标对象。

（2）Active Control 选项

Active Control 选项用于设置 ObjectTarget 游戏对象是否激活。

- 选择 Hide When Not Tracking，只有 3D 物体被跟踪的时候，ObjectTarget 游戏对象才被激活。如果 3D 物体没有被跟踪，则 ObjectTarget 游戏对象不被激活。默认为该选项。
- 选择 Hide Before First Found，当 3D 物体第一次被跟踪以后，ObjectTarget 游戏对象就被激活，之后一直处于激活状态。
- 选择 None，ObjectTarget 游戏对象始终被激活。

当一个 ObjectTarget 游戏对象被激活但是又没有被跟踪的时候，其位置和角度不会变化。

（3）Source Type 选项

该选项用于设置跟踪类型，通常情况下使用 Object File 即可。Target 只是在程序控制的时候会使用到。

（4）事件

ObjectTargetController 类提供了 4 个事件，分别是 3D 物体被识别（TargetFound）、被识别的 3D 物体从视野消失（TargetLost）、3D 物体加载完成（TargetLoad）和 3D 物体卸载完成（TargetUnload）。通过订阅这些事件就能实现对应的程序控制。

```
controller.TargetFound += () =>
{
    ...
};
controller.TargetLost += () =>
{
    ...
};
controller.TargetLoad += (Target target, bool status) =>
{
    ...
};
controller.TargetUnload += (Target target, bool status) =>
{
    ...
};
```

2.4.2　跟踪 3D 物体

1. 导入模型相关内容

将 hexagon.obj、hexagon.mtl 和 hexagon.jpg 文件拖到 StreamingAssets 目录下，作为跟踪用的内容，如图 2-61 所示。

将 hexagon.obj 文件拖到 EasyAR 4Learn/Models 目录下，作为识别后显示用的模型，如图 2-62 所示。

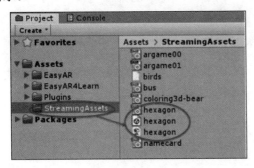
图 2-61

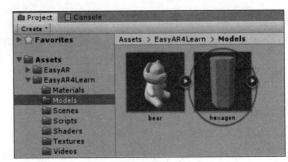
图 2-62

将 hexagon.jpg 文件拖到 EasyAR 4Learn/Textures 目录下，作为模型纹理，如图 2-63 所示。

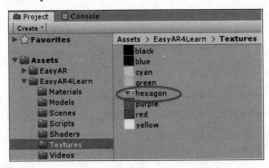
图 2-63

将 hexagon_hexagonMat.mat 文件拖到 EasyAR 4Learn/Materials 目录下，作为材质文件，并设置材质的纹理为 hexagon.jpg，如图 2-64 所示。

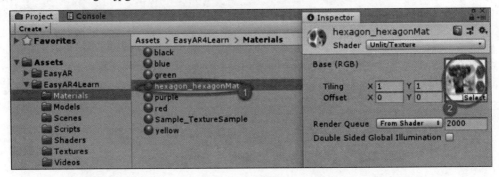
图 2-64

2. 添加并设置基础内容

在目录 EasyAR 4Learn/Scenes 下新建一个场景并命名为 ObjectTarget，设置 Main Camera 游戏对象清除标志 Clear Flags 为 Solid Color。

将 EasyAR/Prefabs/Composites 目录下的 EasyAR_ObjectTracker-1_CloudRecognizer-1 预制件拖到场景中，如图 2-65 所示。

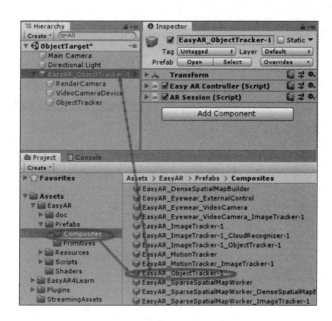

图 2-65

3. 设置跟踪对象

将 EasyAR/Prefabs/Primitives 目录下的 ObjectTarget 预制件拖到场景中。设置 Obj Path 属性为 hexagon.obj，即要跟踪的.obj 文件的路径；修改 Extra File Paths 的 Size 值为"2"，并添加另外 2 个文件的路径，即 hexagon.mtl 和 hexagon.jpg，设置 Name 属性和 Scale 属性，如图 2-66 所示。

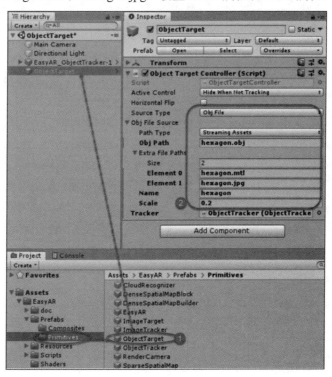

图 2-66

4. 添加并设置显示的模型

将hexagon模型拖到ObjectTarget游戏对象下作为跟踪后显示的模型；修改Rotation的值为"0，180，0"，选择模型以保持方向一致，如图2-67所示。

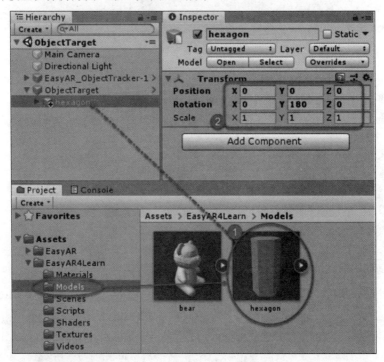

图 2-67

选中 hexagon:hexagon 游戏对象，设置其材质是导入的材质，如图 2-68 所示。

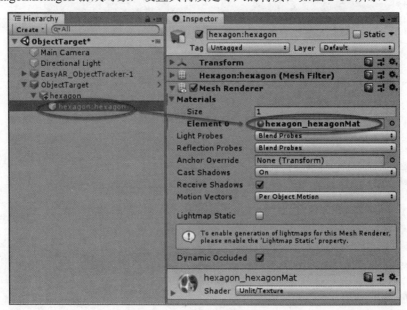

图 2-68

此时运行，当识别到 3D 物体以后就会显示模型，模型正好覆盖原有物体，直观上就是黑白的变成彩色的了，如图 2-69 所示。

图 2-69

5. 添加其他的模型显示

在 ObjectTarget 游戏对象下添加一个方块，修改位置和大小，如图 2-70 所示。

图 2-70

运行，新添加的方块也会一起显示，如图 2-71 所示。

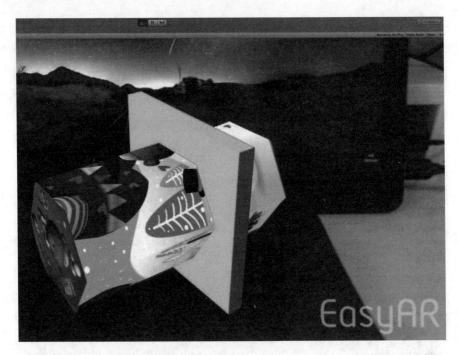

图 2-71

物体跟踪也可以同时跟踪多个物体。跟踪多个物体的逻辑和平面图像跟踪多个图像是一样的，包括程序的使用，只是参数类型有所改变，方法是一样的。考虑到物体跟踪使用得很少，这里就不说明了。

第 3 章
◀ 空间相关内容 ▶

空间相关的内容（即表面跟踪、运动跟踪、稀疏空间地图和稠密空间地图）无法直接在 Unity 编辑器里面调试，必须打包发布成移动应用，安装在设备上以后才能正常运行。直接在编辑器运行会提示 VIOCameraDevice not available 或者 easyar.SurfaceTracker not available，如图 3-1 所示。

图 3-1

如果是第一次使用 Unity，务必先编译发布一个空的项目，确保 Unity 环境的配置正确。

3.1 表面跟踪和运动跟踪

3.1.1 总体说明

表面跟踪和运动跟踪类似，其目的都是通过感知设备在现实空间的变化，来实现将 3D 模型显示得像是位于现实空间的某个位置。

实际使用的方式也类似，只要将物体放置在"WorldRoot"游戏对象下成为其子游戏对象，就能自动实现跟踪效果。

表面跟踪对设备的要求更低，支持的设备更多，但是效果更差一些，而且场景中只能放置一个虚拟物体。能实际应用的场景其实不多。

运动跟踪对设备有要求，官方给出了设备支持列表（https://help.easyar.cn/EasyAR%20Sense/v4/Guides/EasyAR-Motion-Tracking-Supported-Devices.html）。这个设备支持列表里面没有提具体的苹果设备，只提到凡是支持 ARKit 的苹果设备都支持 EasyAR 的运动跟踪。

尽管运动跟踪对设备要求比较高，但是能实现的跟踪效果好很多，而且没有虚拟物体限制。实际情况下，更推荐使用运动跟踪而不是表面跟踪。

表面跟踪的基本结构如图 3-2 所示。

运动跟踪的基本结构如图 3-3 所示。

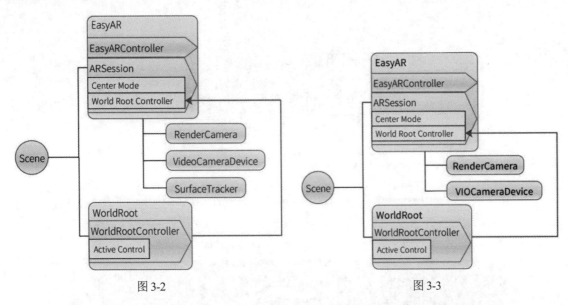

图 3-2 　　　　　　　　　　　　　图 3-3

两者使用的时候都是要设置 Center Mode 属性为 World Root，并将 WorldRoot 游戏对象设置为 World Root Controller 属性的值，同时将需要跟踪的游戏对象放置在 WorldRoot 游戏对象下面即可。

3.1.2 表面跟踪

1. 添加并设置基础内容

在目录 EasyAR 4Learn/Scenes 下新建一个场景并命名为 SurfaceTracker，设置场景中 Main Camera 的 Clear Flags 属性为 Solid Color。将 EasyAR/Prefabs/Composites 目录下的 EasyAR_SurfaceTracker 预制件拖到场景中，如图 3-4 所示。

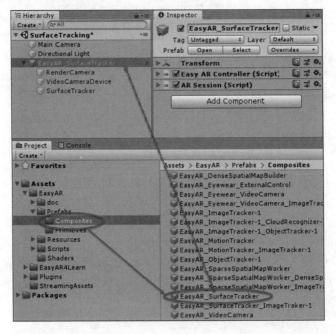

图 3-4

将 EasyAR/Prefabs/Primitives 目录下的 WorldRoot 预制件拖到场景中；选中 EasyAR_SurfaceTracker 游戏对象，将 WorldRoot 游戏对象拖到 World Root Controller 属性中，如图 3-5 所示。

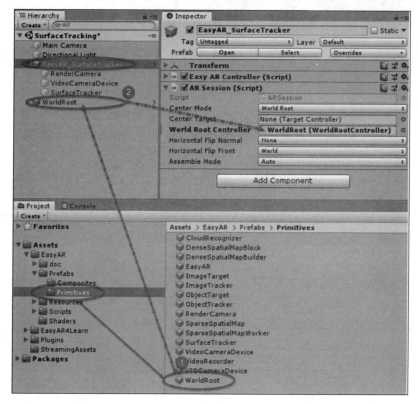

图 3-5

2. 添加虚拟内容

为了显示跟踪效果，在 WorldRoot 游戏对象下添加一个方块。

选中 WorldRoot 游戏对象，单击鼠标右键，在弹出的菜单中选择 3D Object→Cube，添加一个方块，如图 3-6 所示。

设置方块的位置在原点，将 EasyAR4Learn/Textures 目录下的颜色拖到方块上成为其纹理，这样模型就可以通过颜色来区分了，如图 3-7 所示。

在场景根目录下再添加一个方块，设置其位置也是在原点，和 WorldRoot 游戏对象下的方块位置略微错开。

这里设置的两个方块看上去是叠在一起的，如图 3-8 所示。

两个方块的坐标和缩放对象如表 3-1 所示。

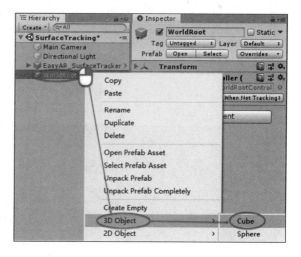

图 3-6

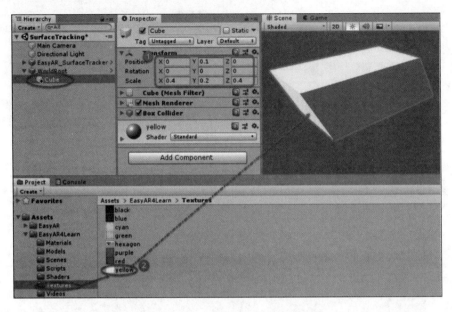

图 3-7

表 3-1 两个方块的坐标和缩放对象

游戏对象	Position	Scale
WorldRoot/Cube	0，0.1，0	0.4，0.2，0.4
Cube	0，-0.1，0	0.4，0.2，0.4

这个例子必须打包后在移动设备上运行。这两个方块（WorldRoot 的原点和场景的原点）会出现在设备前方一点的位置。当设备小范围发生移动旋转的时候，方块位置变化不大。当设备大范围左右移动旋转以后，方块位置也会发生偏移，而且受周围环境影响，但是两个方块的相对位置基本不变，如图 3-9 所示。

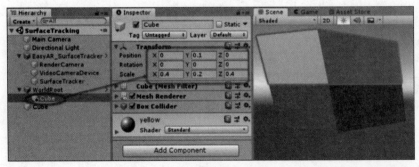

图 3-8

图 3-9

3.1.3 运动跟踪

1. 添加并设置基础内容

在目录 EasyAR 4Learn/Scenes 下新建一个场景并命名为 MotionTracker。设置场景中 Main Camera 的 Clear Flags 属性为 Solid Color。将 EasyAR/Prefabs/Composites 目录下的 EasyAR_MotionTracker 预制件拖到场景中，如图 3-10 所示。

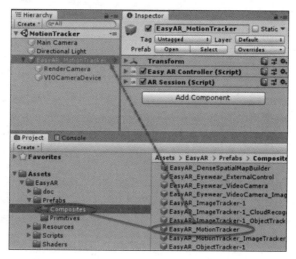

图 3-10

将 EasyAR/Prefabs/Primitives 目录下的 WorldRoot 预制件拖到场景中。选中 EasyAR_MotionTracker 游戏对象，将 WorldRoot 游戏对象拖到 World Root Controller 属性中，如图 3-11 所示。

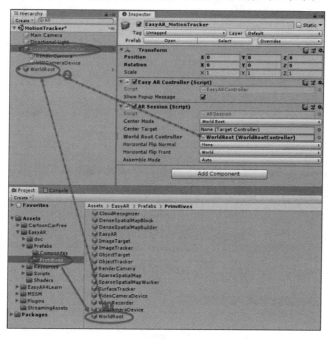

图 3-11

2. 添加虚拟内容

和表面跟踪一样，在中心位置添加两个错开的方块，一个方块在场景根目录下，一个方块在 WorldRoot 游戏对象下。

在 WorldRoot 游戏对象下再添加一些模型，在三个轴的正方向和负方向各添加一个模型并用颜色区分，如图 3-12 所示。

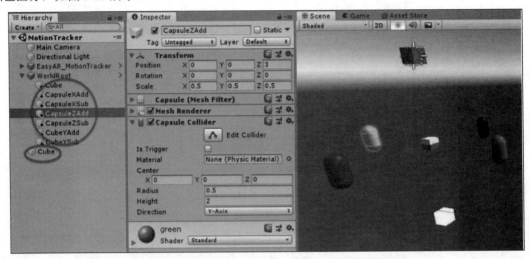

图 3-12

WorldRoot 下的游戏对象的坐标和颜色如表 3-2 所示。

表 3-2　WorldRoot 下的游戏对象

WorldRoot 下的游戏对象	Position	颜　　色
CapsuleXAdd	3，0，0	黑
CapsuleXSub	-3，0，0	蓝
CapsuleZAdd	0，0，3	绿
CapsuleZSub	0，0，-3	粉
CubeYAdd	0，3，0	红
CubeYSub	0，-3，0	白

打包后，在移动设备上运行。两个叠在一起的方块（WorldRoot 的原点和场景的原点）出现在设备所在位置。Y 轴正方向永远朝上。如果设备竖着拿（Portrait 模式），运动跟踪开始的时候，屏幕背面是 Z 轴正方向。X 轴和 Z 轴的方向与东南西北无关，只和启动时设备的角度有关。

当设备移动或旋转时，无论是叠在一起的方块的位置还是两者相对的位置角度，基本都不会发生变化。

使用下来感觉表面跟踪和运动跟踪对环境其实是有要求的，可以认为其对环境的要求和稀疏空间地图对环境的要求类似，如图 3-13 所示。

图 3-13

3.2 稀疏空间地图

3.2.1 总体说明

稀疏空间地图可以简单地理解为把平面图像跟踪的技术从二维平面变成三维空间的应用。平面图像跟踪是在图像上标记出一系列的特征点,用于识别并跟踪平面图像。稀疏空间地图是在空间标记出一系列的特征点用于识别并跟踪空间。

稀疏空间地图对应用环境的要求和平面图像识别可以比照理解,周围环境需要足够丰富,不能有大片的单色区域、透明区域。此外,光照、角度都会对建立地图和定位地图产生影响。

官方给出了建立地图和定位地图的建议,地址如下:

https://help.easyar.cn/EasyAR%20Sense/v4/Guides/EasyAR-Sparse-Spatial-Map.html

稀疏空间地图的基础是运动跟踪,所以在场景中首先要有运动跟踪的全套游戏对象,包括设置,主要是 SparseSpatialMapWorker 和 SparseSpatialMap 这两个游戏对象。

稀疏空间地图基本结构如图 3-14 所示。

1. SparseSpatialMapWorker 游戏对象相关

SparseSpatialMapWorker 游戏对象的主要属性如图 3-15 所示。

(1) Localization Mode 属性

该属性在建立地图的时候通常选 Until Success,在定位的时候通常选 Keep Update。Localization Mode 选项如表 3-3 所示。

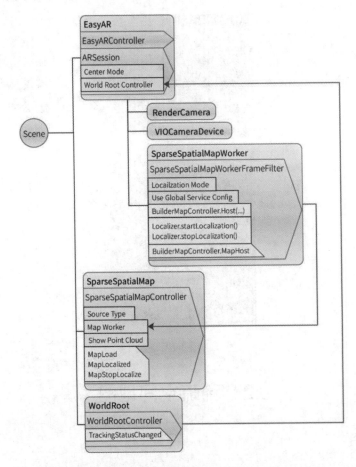

图 3-14

图 3-15

表 3-3 Localization Mode 选项

选 项	值	说 明
Until Success	0	一直尝试定位,一旦定位成功就停止继续尝试
Once	1	尝试定位一次
Keep Update	2	一直尝试定位,即使定位成功,依然继续尝试,并在定位再次成功时调整结果到更准确的位置和姿态
Continous Localize	3	一直尝试定位,仅在定位到新地图时调整结果到对应位置和姿态

（2）Use Global Service Config 选项

通过该选项可以设置使用全局定义的稀疏空间地图信息还是针对场景重新定义和设置，默认为使用全局定义。

（3）BuilderMapController.Host(...)方法

该方法用于保存地图，需要输入的参数是地图的名称和地图的缩略图，但是缩略图的格式处理起来稍微麻烦一点，好在可以输入"null"。

（4）BuilderMapController.MapHost 事件

该事件用于返回地图保存情况，有 3 个参数，是地图保存成功后的信息（包括名称、ID），包括是否保存成功的状态和错误信息。保存地图的时候，如果没有注册这个事件就无法知道是否保存成功。

官方虽然提供了在程序中保存地图的方法，但是没有提供删除的方法。删除只能通过浏览器登录后台来删除。

官方暂时没有提供任何修改地图的方法。

（5）Localizer.startLocalization()和 Localizer.stopLocalization()方法

这两个方法用来启动和停止本地稀疏空间定位。如果 SparseSpatialMap 游戏对象设置了地图的 ID 和名称，默认会自动启动地图定位。

2. SparseSpatialMap 游戏对象相关

SparseSpatialMap 游戏对象是稀疏空间地图在 Unity 中的载体，每个稀疏空间地图在定位的时候对应一个 SparseSpatialMap 游戏对象。同一个场景可以有多个稀疏空间地图，即多个 SparseSpatialMap 游戏对象。对于希望在某个稀疏空间地图中放置的虚拟物体，将对应的游戏对象放置到对应的 SparseSpatialMap 游戏对象下成为子游戏对象即可，如图 3-16 所示。

图 3-16

（1）Source Type 属性

该属性用于设置稀疏空间地图的作用，即是用于建立地图 Map Builder 还是用于定位 Map Manager。其中，选择了 Map Manager 以后，可以输入地图 ID 和名称来实现定位。

（2）Map Worker 属性

该属性必须关联对应的 SparseSpatialMapWorker 游戏对象，通常不需要设置。

（3）Show Point Cloud 选项

该选项可以设置是否点云的效果。在建图的时候，显示点云的效果能帮助使用者更好地建立稀疏空间地图。

EasyAR 也提供了自己定义点云效果的方式，使用起来很方便。

（4）MapLoad、MapLocalized、MapStopLocalize 事件

这 3 个事件都是在地图定位的时候需要用到的。当启动了稀疏空间地图本地化之后，首先会从服务器下载地图信息。此时会触发 MapLoad 事件，该事件会反馈是否下载到了地图及相关信息。当本地获取到地图信息以后，会开始利用地图进行定位，当定位成功以后会触发 MapLocalized 事件。在地图定位成功之后，如果发生无法定位的情况，比如离开了定位的区域、遮挡了摄像头，就会触发 MapStopLocalize 事件。其中，MapLocalized 事件可以触发多次。

3.2.2 建立地图

1. 添加并设置基础内容

在 EasyAR 4Learn/Scenes 目录下新建一个场景并命名为 BuildMap。设置场景中的 Main Camera 的 Clear Flags 属性为 Solid Color。将 EasyAR/Prefabs/Composites 目录下的 EasyAR_SparseSpatialMapWorker 预制件拖到场景中，如图 3-17 所示。

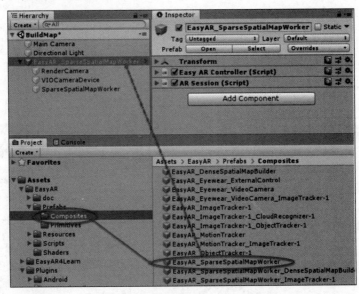

图 3-17

将 EasyAR/Prefabs/Primitives 目录下的 WorldRoot 预制件拖到场景中；选中 EasyAR_SparseSpatialMapWorker 游戏对象，将 WorldRoot 游戏对象拖到 World Root Controller 属性中，如图 3-18 所示。

将 EasyAR/Prefabs/Primitives 目录下的 SparseSpatialMap 预制件拖到场景中，如图 3-19 所示。

2. 添加辅助模型

像运动跟踪中介绍的那样，在根目录添加方块，在 WorldRoot 游戏对象下添加模型。

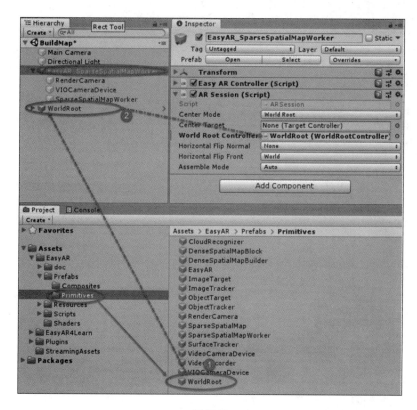

图 3-18

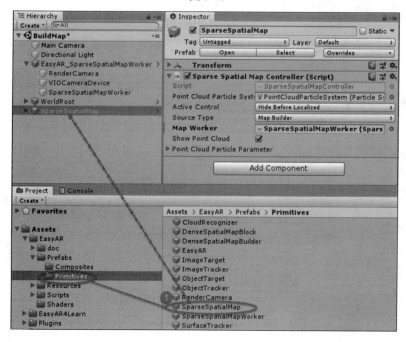

图 3-19

在 SparseSpatialMap 游戏对象下添加用于分辨方向的模型,用形状来和 WorldRoot 游戏对象下的模型进行区分,如图 3-20 所示。

Unity3D 平台 AR 开发快速上手：基于 EasyAR 4.0

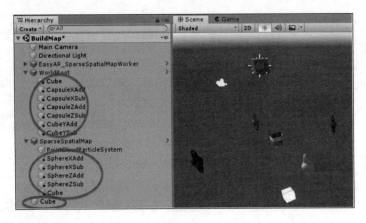

图 3-20

SparseSpatialMap 下新增游戏对象的坐标和颜色如表 3-4 所示。

表 3-4 复制新增游戏对象的坐标和颜色

SparseSpatialMap 下的游戏对象	坐　　标	颜　　色
SphereXAdd	3，0，0	黑
SphereXSub	-3，0，0	蓝
SphereZAdd	0，0，3	绿
SphereZSub	0，0，-3	粉
Cube	0，0.3，0	蓝

3. 添加脚本相关内容

在场景中添加一个按钮 Button，设置在场景右下角，用于保存建立的地图。再添加一个面板 Panel 在左下方，并在面板 Panel 下添加一个文本 Text 作为子游戏对象，用于显示提示信息。添加面板的目的是避免文字看不清楚。

新建一个脚本 BuildMapController 并将其拖到一个新建的空的游戏对象上，如图 3-21 所示。

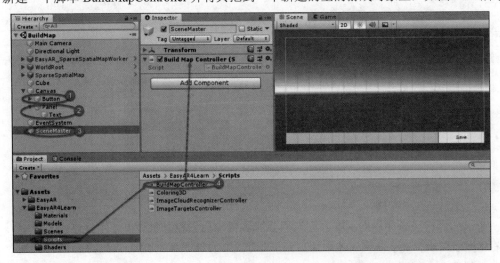

图 3-21

4. 脚本内容说明

在保存的方法中注册事件，然后调用 Host 方法即可。

```
private void Save()
{
   btnSave.interactable = false;
   //注册地图保存结果反馈事件
   mapWorker.BuilderMapController.MapHost += SaveMapHostBack;
   //保存地图
   try
   {
      //保存地图
      mapWorker.BuilderMapController.Host("LearnMap"..., null);
      text.text = "开始保存地图，请稍等。";
   }
   catch (Exception ex)
   {
      btnSave.interactable = true;
      text.text = "保存出错：" + ex.Message;
   }
}
```

保存方法反馈事件中，通过 MapInfo 可以获取到已经保存的地图的 ID 和名称。

```
private void SaveMapHostBack(
SparseSpatialMapController.SparseSpatialMapInfo mapInfo,
bool isSuccess, string error)
{
   if (isSuccess)
   {
      PlayerPrefs.SetString("MapID", mapInfo.ID);
      PlayerPrefs.SetString("MapName", mapInfo.Name);
      text.text = "地图保存成功。\r\nMapID: " ...;
   }
   else
   {
      btnSave.interactable = true;
      text.text = "地图保存出错：" + error;
   }
}
```

完整代码如下：

```
void Start()
{
   //稀疏空间地图初始
   session = FindObjectOfType<ARSession>();
   mapWorker = FindObjectOfType<SparseSpatialMapWorkerFrameFilter>();
```

```csharp
        map = FindObjectOfType<SparseSpatialMapController>();
        //注册追踪状态变化事件
        session.WorldRootController.TrackingStatusChanged +=
    OnTrackingStatusChanged;
        //初始化保存按钮
        btnSave = GameObject.Find("/Canvas/Button").GetComponent<Button>();
        btnSave.onClick.AddListener(Save);
        btnSave.interactable = false;
        if (session.WorldRootController.TrackingStatus ==
MotionTrackingStatus.Tracking)
        {
            btnSave.interactable = true;
        }
        else
        {
            btnSave.interactable = false;
        }
        //初始化显示文本
        text = GameObject.Find("/Canvas/Panel/Text").GetComponent<Text>();
    }
    private void Save()
    {
        btnSave.interactable = false;
        //注册地图保存结果反馈事件
        mapWorker.BuilderMapController.MapHost += SaveMapHostBack;
        //保存地图
        try
        {
            //保存地图
            mapWorker.BuilderMapController.Host("LearnMap" + ..., null);
            text.text = "开始保存地图,请稍等。";
        }
        catch (Exception ex)
        {
            btnSave.interactable = true;
            text.text = "保存出错:" + ex.Message;
        }
    }
    private void SaveMapHostBack(
    SparseSpatialMapController.SparseSpatialMapInfo mapInfo, bool isSuccess, string
error)
    {
        if (isSuccess)
        {
            PlayerPrefs.SetString("MapID", mapInfo.ID);
            PlayerPrefs.SetString("MapName", mapInfo.Name);
            text.text = "地图保存成功。\r\nMapID: " +...;
```

```
            }
            else
            {
                btnSave.interactable = true;
                text.text = "地图保存出错：" + error;
            }
        }
        private void OnTrackingStatusChanged(MotionTrackingStatus status)
        {
            if (status == MotionTrackingStatus.Tracking)
            {
                btnSave.interactable = true;
                text.text = "进入跟踪状态。";
            }
            else
            {
                btnSave.interactable = false;
                text.text = "退出跟踪状态。" + status.ToString();
            }
        }
```

打包以后在设备上运行。进入以后，会显示点云效果。稀疏空间地图 SparseSpatialMap 游戏对象和运动跟踪 WorldRoot 游戏对象的原点方向都是一致的，如图 3-22、图 3-23 所示。当扫描完周围空间以后，单击保存按钮，就可以将地图保存到服务器。在提示文本中会显示地图的 ID 和名称，如图 3-24 所示。

图 3-22

图 3-23

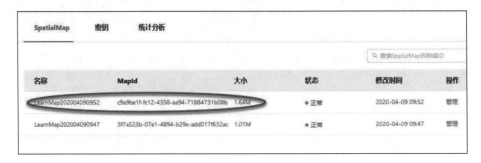

图 3-24

3.2.3 本地化稀疏空间地图

1. 添加并设置场景基础内容

打开建立地图的场景，选中场景名称，单击鼠标右键，在弹出的菜单中选择 Save Scene As，将场景另存为 LocalizeMap 作为本地化的场景，如图 3-25 所示。

打开 LocalizeMap 场景，选中 SparseSpatialMapWorker 游戏对象，修改 Localization Mode 属性为 Keep Update，如图 3-26 所示。

选中 SparseSpatialMap 游戏对象，修改 Source Type 属性为 Map Manager，如图 3-27 所示。

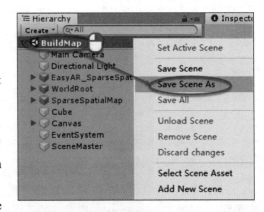

图 3-25

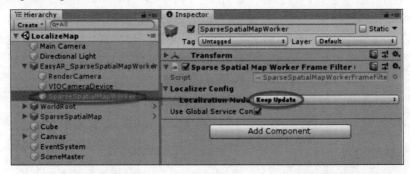

图 3-26

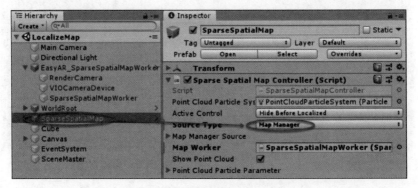

图 3-27

2. 修改界面

删除原有界面 UI。单击菜单 GameObject→UI，选中下面的 Text、Button、Input Field，为界面添加文本显示、按钮和输入框，如图 3-28 所示。

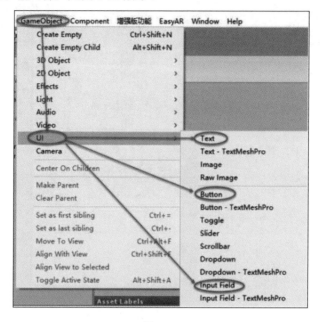

图 3-28

设置界面里有 2 个输入框，分别用于输入地图 ID 和地图名称；有一个文本框，用于显示提示内容；有 2 个按钮，用于本地化地图和停止定位，如图 3-29 所示。

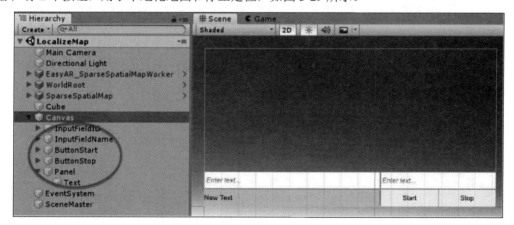

图 3-29

3. 添加脚本内容

选中脚本所在游戏对象，删除原有脚本。新建一个脚本 LocalizeMapController 并拖到游戏对象上成为其组件，如图 3-30 所示。

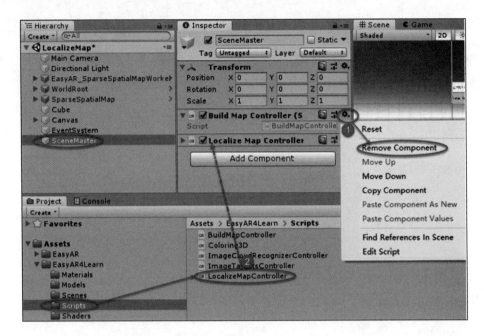

图 3-30

在 Start 事件中注册地图本地化相关的事件。

```
void Start()
{
   ...
   map.MapLoad += MapLoadBack;     //注册地图加载事件
   map.MapLocalized += LocalizedMap;   //注册定位成功事件
   map.MapStopLocalize += StopLocalizeMap; //注册停止定位事件
   ...
   StartLocalization();
}
```

本地化使用输入框中的 ID 和名称进行。

```
public void StartLocalization()
{
   if (inputID.text.Length > 0 && inputName.text.Length > 0)
   {
      map.MapManagerSource.ID = inputID.text;
      map.MapManagerSource.Name = inputName.text;
   }
   text.text = "开始本地化地图。";
   mapWorker.Localizer.startLocalization();
}
public void StopLocalization()
{
   mapWorker.Localizer.stopLocalization();
}
```

完整代码如下：

```csharp
void Start()
{
    //稀疏空间地图初始
    session = FindObjectOfType<ARSession>();
    mapWorker = FindObjectOfType<SparseSpatialMapWorkerFrameFilter>();
    map = FindObjectOfType<SparseSpatialMapController>();

    //如果之前有建立过地图且文本框没有预设值
    if (inputID.text.Length <= 0)
    {
        inputID.text = PlayerPrefs.GetString("MapID", "");
        inputName.text = PlayerPrefs.GetString("MapName", "");
    }

    map.MapLoad += MapLoadBack;  //注册地图加载事件
    map.MapLocalized += LocalizedMap;    //注册定位成功事件
    map.MapStopLocalize += StopLocalizeMap; //注册停止定位事件

    StartLocalization();
}
private void MapLoadBack(
SparseSpatialMapController.SparseSpatialMapInfo mapInfo, bool isSuccess, string error)
{
    if (isSuccess)
    {
        text.text = "地图" + mapInfo.Name + "加载成功。";
    }
    else
    {
        text.text = "地图加载失败。" + error;
    }
}
private void LocalizedMap()
{
    text.text = "稀疏空间地图定位成功。" + DateTime.Now.ToShortTimeString();
}
private void StopLocalizeMap()
{
    text.text = "稀疏空间地图停止定位。" + DateTime.Now.ToShortTimeString();
}
public void StartLocalization()
{
    //文本框内容不为空
    if (inputID.text.Length > 0 && inputName.text.Length > 0)
    {
```

```
        map.MapManagerSource.ID = inputID.text;
        map.MapManagerSource.Name = inputName.text;
    }
    text.text = "开始本地化地图。";
    mapWorker.Localizer.startLocalization();
}
public void StopLocalization()
{
    mapWorker.Localizer.stopLocalization();
}
```

4. 设置脚本

选中 ButtonStart 游戏对象，单击 On Click 标签下的"+"按钮，添加 On Click 事件；将 SceneMaster 游戏对象拖到 On Click 下的框中，设置事件响应的方法来自 SceneMaster 游戏对象；选择右边的下拉菜单，选择 LocalizeMapController 下的 StartLocalization，即设置事件响应的方法是 LocalizeMapController 脚本下的 StartLocalization 方法，如图 3-31 所示。

用同样的方法设置 ButtonStop 游戏对象的 On Click 事件响应的方法，把它设为 SceneMaster 游戏对象下 LocalizeMapController 组件的 StopLocalization 方法。

打包后在设备上运行，默认获取之前建立的地图。

单击 Start 按钮开始本地化地图，单击 Stop 按钮停止定位。停止定位以后，输入新的地图 ID 和名称，再单击 Start 按钮开始本地化地图。

地图的原点和方向（SparseSpatialMap 游戏对象的位置方向）和本地化时设备的状态无关，和建立地图时的状态一致，如图 3-32 所示。

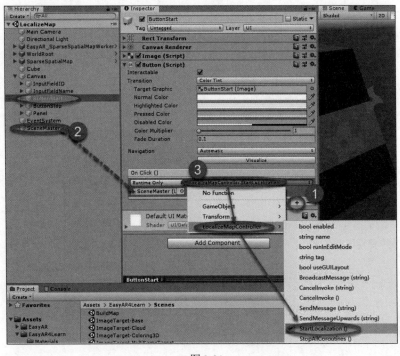

图 3-31

图 3-32

3.3 稠密空间地图

3.3.1 总体说明

稀疏空间地图的主要作用是定位，稠密空间地图的主要作用是重建。利用 RGB 相机图像对周围环境进行三维稠密重建，得到稠密的点云地图和网格地图，再利用网络地图对虚拟物体实现遮挡和碰撞。稠密空间地图官方没有提供持久化的方法。另外，稠密空间地图不仅对设备的传感器有要求，还对计算能力有要求。不过现在能支持运动跟踪的设备都是高端设备，计算能力上应该问题不大。

稠密空间地图结构很简单，在 DenseSpatialMapBuilder 游戏对象下有一个叫 DenseSpatialMapBuilderFrameFilter 的脚本处理网络地图。

稠密空间地图基本结构如图 3-33 所示。

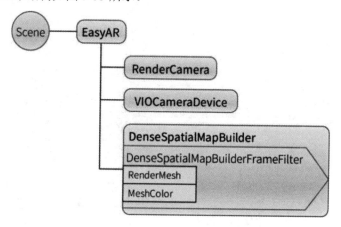

图 3-33

- RenderMesh 属性：当该属性为 true 时，能够显示网格对虚拟物体的遮挡效果。
- MeshColor 属性：该属性是网格地图的颜色，可以设置为具体颜色，也可以设置为完全透明，看上去像是被实际物体遮挡的效果。

3.3.2 建立并使用稠密空间地图

1. 添加并设置基础内容

在 EasyAR 4Learn/Scenes 目录下新建一个场景并命名为 Dense。设置场景中 Main Camera 的 Clear Flags 属性为 Solid Color。将 EasyAR/Prefabs/Composites 目录下的 EasyAR_DenseSpatialMapBuilder 预制件拖到场景中，如图 3-34 所示。

单击菜单 GameObject→UI→Toggle，为场景添加 2 个切换按钮，分别命名为 ToggleRender 和 ToggleTransparent。设置 2 个切换按钮的大小和位置。这里是设置在平面左下角，如图 3-35 所示。

新建一个空的游戏对象并命名为 SceneMaster。在目录 EasyAR 4Learn/Scripts 下新建脚本 DenseController 并拖到 SceneMaster 目录下成为其组件，如图 3-36 所示。

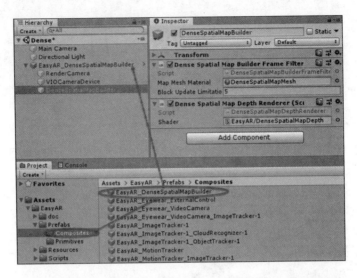

图 3-34

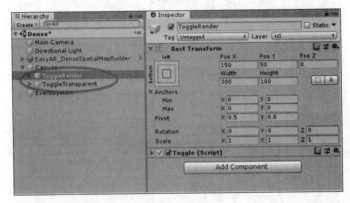

图 3-35

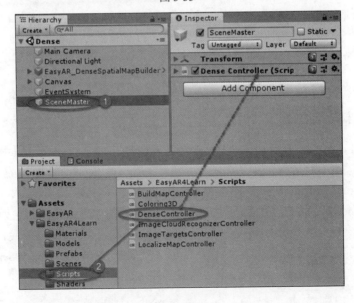

图 3-36

2. 设置生成的球体

单击菜单 GameObject→3D Object→Sphere，在场景中添加一个球体；修改球体名称为 BigBall；设置球体位置为场景原点，球体半径为 0.5（米）；将 EasyAR 4Learn/Textures 目录下的贴图拖到球体上改变颜色，使其更显眼，如图 3-37 所示。

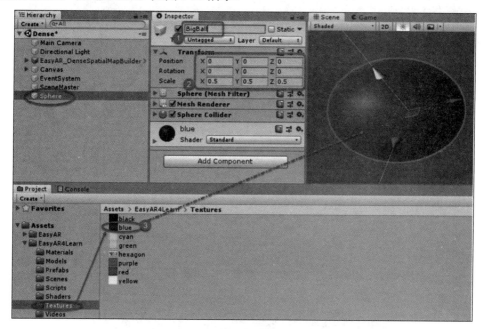

图 3-37

选中球体，单击菜单 Component→Physics→Rigidbody，为球体添加刚体组件（默认即可），使其能够实现碰撞，如图 3-38 所示。

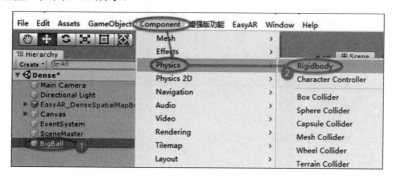

图 3-38

将 BigBall 游戏对象拖到 EasyAR 4Learn/Prefabs 目录下成为预制件，如图 3-39 所示。

3. 编写脚本

在 Update 方法中添加单击屏幕后端处理方法。

在屏幕只有一个单击且未发生 UI 击穿的情况下，在屏幕单击的位置添加一个球体，并给一个向前的力。

```
if (Input.GetMouseButtonDown(0) &&
 Input.touchCount > 0 &&
 !EventSystem.current.IsPointerOverGameObject(Input.GetTouch(0).fingerId))
{
    Ray ray = Camera.main.ScreenPointToRay(Input.touches[0].position);
    var launchPoint = Camera.main.transform;
    var ball = Instantiate(prefab, launchPoint.position, launchPoint.rotation);
    var rigid = ball.GetComponent<Rigidbody>();
    rigid.velocity = Vector3.zero;
    rigid.AddForce(ray.direction * 15f + Vector3.up * 5f);
}
```

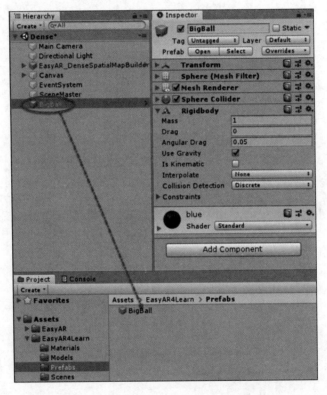

图 3-39

添加设置 RenderMesh 和 MeshColor 的方法。

```
public void RenderMesh(bool show)
{
    ...
    dense.RenderMesh = show;
}
public void TransparentMesh(bool trans)
{
    ...
    if (trans)
    {
```

```
            dense.MeshColor = Color.gray;
        }
        else
        {
            dense.MeshColor = Color.clear;
        }
    }
```

完整的脚本内容如下：

```
public class DenseController : MonoBehaviour
{
    public GameObject prefab;
    public DenseSpatialMapBuilderFrameFilter dense;

    void Start()
    {
        dense.MeshColor = Color.gray;
    }

    void Update()
    {
        if (Input.GetMouseButtonDown(0) &&
         Input.touchCount > 0 &&
         !EventSystem.current.IsPointerOverGameObject(Input.GetTouch(0).fingerId))
        {
            Ray ray = Camera.main.ScreenPointToRay(Input.touches[0].position);
            var launchPoint = Camera.main.transform;
            var ball = Instantiate(prefab, launchPoint.position, launchPoint.rotation);
            var rigid = ball.GetComponent<Rigidbody>();
            rigid.velocity = Vector3.zero;
            rigid.AddForce(ray.direction * 15f + Vector3.up * 5f);
        }
    }

    public void RenderMesh(bool show)
    {
        if (!dense)
        {
            return;
        }
        dense.RenderMesh = show;
    }

    public void TransparentMesh(bool trans)
    {
```

```
        if (!dense)
        {
            return;
        }
        if (trans)
        {
            dense.MeshColor = Color.gray;
        }
        else
        {
            dense.MeshColor = Color.clear;
        }
    }
}
```

4．设置脚本

选中SceneMaster游戏对象，将DenseSpatialMapBuilder游戏对象拖到Dense属性中为其赋值。

将EasyAR 4Learn/Prefabs目录下的BigBall游戏对象拖到Prefab属性中为其赋值，如图3-40所示。

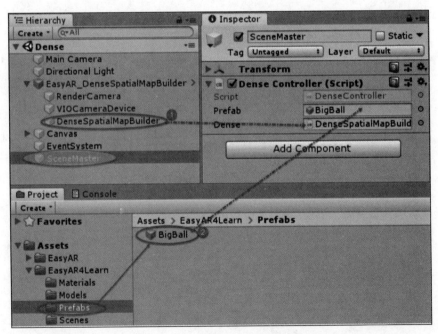

图 3-40

选中ToggleRender游戏对象，单击On Value Changed标签下的"+"按钮，添加On Value Changed事件；将SceneMaster游戏对象拖到On Value Changed下的框中，设置事件响应的方法来自SceneMaster游戏对象；单击右边的下拉按钮，选择DenseController下的RenderMesh，即设置事件响应的方法是DenseController脚本下的RenderMesh方法，如图3-41所示。

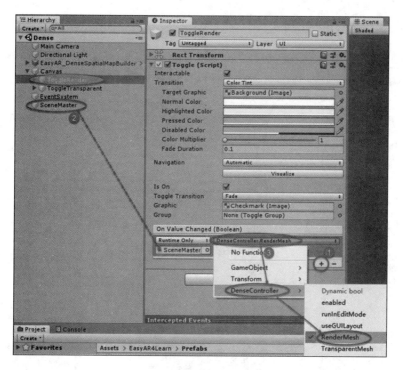

图 3-41

用同样的方法设置 ToggleTransparent 切换按钮的 On Value Changed 事件的响应方法，把它设置为 DenseController 脚本下的 TransparentMesh 方法，如图 3-42 所示。

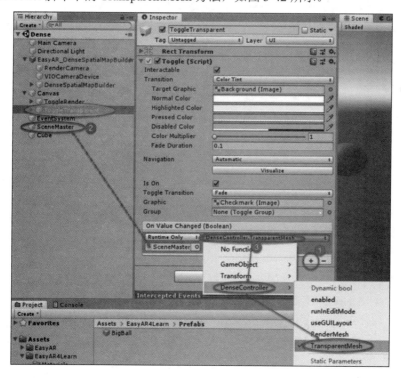

图 3-42

打包后在设备上的运行效果如下：

- 当 Render 和 Transparent 按钮都选中的时候，会显示遮挡效果以及网格地图，如图 3-43 所示。
- 去掉 Transparent 以后，则不会显示网格地图，如图 3-44 所示。
- 去掉 Render 以后，遮挡效果消失，如图 3-45 所示。

图 3-43　　　　　　　　图 3-44　　　　　　　　图 3-45

第 4 章
◀ 屏 幕 录 像 ▶

4.1 总体说明

屏幕录像不是增强现实的功能，只是 EasyAR SDK 自带的一个功能。该功能限制颇多，只能在移动设备上使用，而且没有办法直接录制 UI 界面。屏幕录像功能本质上录的是 RenderTexture，主要是在基础结构上添加了一个 VideoRecorder 游戏对象。另外，需要动态地往 Main Camera 主摄像机游戏对象上添加 CameraRecorder 脚本。

屏幕录像的基本结构如图 4-1 所示。

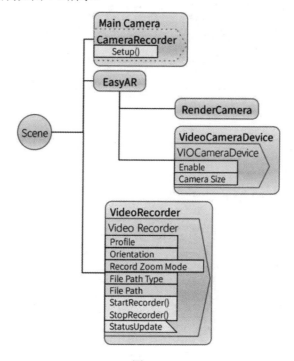

图 4-1

4.1.1 CameraRecorder 脚本相关

CameraRecorder 脚本需要动态地添加到 Main Camera 主摄像机游戏对象上，该脚本可以通过 Setup 方法设置录像内容的水印。

4.1.2 VideoCameraDevice 游戏对象相关

（1）Enable 属性

该属性可以用于关闭摄像头内容，这样就能只录制屏幕内容。

（2）Camera Size 属性

该属性用于设置摄像头获取的视频的分辨率。

4.1.3 VideoRecorder 游戏对象相关

（1）Profile 属性

该属性用于设置录制效果，最高可以录制 1080P 的高清内容，如图 4-2 所示。

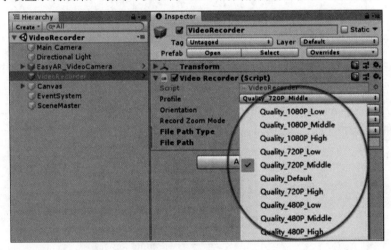

图 4-2

（2）Orientation 属性

该属性用于设置录像时是横屏（Landscape）还是竖屏（Portrait），或者根据当前屏幕情况设定（Screen Adaptive），如图 4-3 所示。

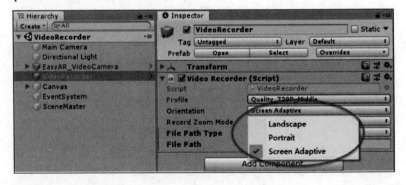

图 4-3

（3）Record Zoom Mode 属性

该属性用于设置屏幕剪裁，如图 4-4 所示。

图 4-4

- No Zoom And Clip：如果输出宽高比与输入不符，内容会被剪裁到适合输出的比例。
- Zoom In With All Content：如果输出宽高比与输入不符，内容将不会被剪裁，在某个维度上会有黑边。

（4）File Path Type 属性

该属性设置文件的路径类型（见图 4-5），Absolute 是绝对路径，Persistent Data Path 是相对于持久数据路径的相对路径。通常使用后者，因为该路径下能保证可写入。

图 4-5

（5）File Path 属性

该属性用于设置录制后保存的视频文件的具体路径及文件名，如图 4-6 所示。

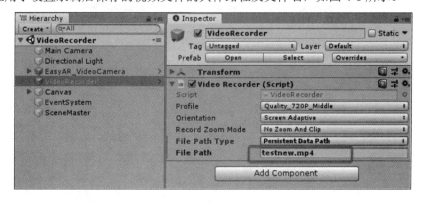

图 4-6

4.1.4 禁用多线程渲染

单击菜单 File→Build Settings，在 Scenes In Build 窗口中单击 Player Settings，在 Other Settings 面板下取消 Multithreaded Rendering 选项，如图 4-7 所示。

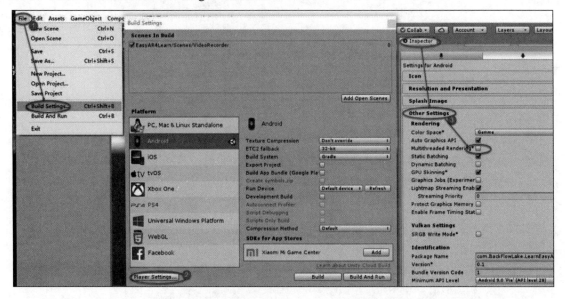

图 4-7

4.2 使用屏幕录像功能

1. 导入脚本

只导入 EasyAR 4.0 SDK，要使用屏幕录像功能还比较麻烦，还需要将官方示例中的一个脚本导入过来，这样使用起来会简单很多，如图 4-8 所示。

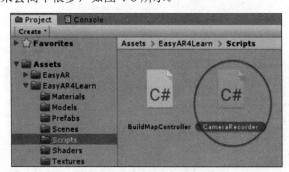

图 4-8

2. 添加并设置基础内容

在 EasyAR 4Learn/Scenes 目录下新建一个场景并命名为 VideoRecorder。设置场景中的 Main Camera 的 Clear Flags 属性为 Solid Color。

将 EasyAR/Prefabs/Composites 目录下的 EasyAR_VideoCamera 预制件拖到场景中，如图 4-9 所示。

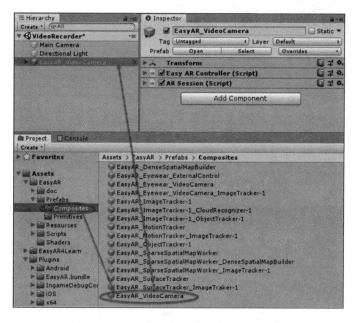

图 4-9

将 EasyAR/Prefabs/Primitives 目录下的 VideoRecorder 预制件拖到场景中，设置 VideoreCorder 属性，设置 File Path Type 为 Persistent Data Path，设置 File Path 为具体文件名，如图 4-10 所示。

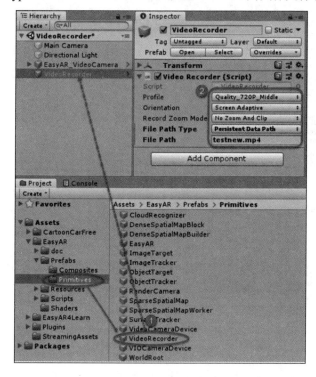

图 4-10

单击菜单 GameObject→UI→Button 往场景中添加按钮，用于实现开始录像和停止录像功能。添加 2 个按钮，设置名称和位置；单击菜单 GameObject→UI→Text，往场景中添加一个显示信息的文本框，如图 4-11 所示。

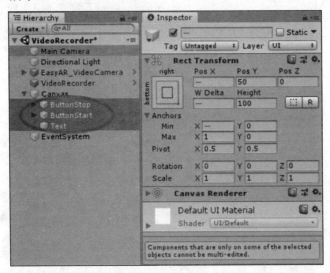

图 4-11

在场景中新建一个空的游戏对象并命名为 SceneMaster；在 EasyAR 4Learn/Scripts 目录下添加一个脚本 RecorderController，并拖到 SceneMaster 游戏对象下成为其脚本，如图 4-12 所示。

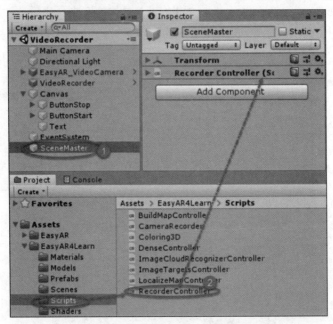

图 4-12

3．编写脚本内容

在脚本的 Awake 事件中，添加对 StatusUpdate 事件的侦听内容，将获取到的信息显示到文本框。

```
videoRecorder.StatusUpdate += (status, msg) =>
{
    if (status == RecordStatus.OnStarted)
    {
        uiText.text = "Recording start";
    }
    if (status == RecordStatus.FailedToStart ||
        status == RecordStatus.FileFailed || status == RecordStatus.LogError)
    {
        uiText.text = "Recording Error: " + status + ", details: " + msg;
    }
    Debug.Log("RecordStatus: " + status + ", details: " + msg);
};
```

添加开始录制的方法。在运行了 StartRecording 事件后,还需要动态地往主摄像机上添加 CameraRecorder 脚本,并运行该脚本的 Setup 方法。

```
public void StartRecorder()
{
    videoRecorder.StartRecording();
    cameraRecorder = Camera.main.gameObject.AddComponent<CameraRecorder>();
    cameraRecorder.Setup(videoRecorder, null);
}
```

添加停止录像的方法,停止后还需要删除动态添加的脚本。

```
public void StopRecorder()
{
    if (videoRecorder.StopRecording())
    {
        uiText.text = "Recording stop " + videoRecorder.FilePath;
    }
    else
    {
        uiText.text = "Recording failed";
    }
    if (cameraRecorder)
    {
        cameraRecorder.Destroy();
    }
}
```

完整的脚本内容如下:

```
using UnityEngine;
using UnityEngine.UI;
using easyar;
using VideoRecording;
```

```csharp
public class RecorderController : MonoBehaviour
{
    public Text uiText;
    public VideoRecorder videoRecorder;
    private CameraRecorder cameraRecorder;

    private void Awake()
    {
        videoRecorder.StatusUpdate += (status, msg) =>
        {
            if (status == RecordStatus.OnStarted)
            {
                uiText.text = "Recording start";
            }
            if (status == RecordStatus.FailedToStart ||
            status == RecordStatus.FileFailed || status == RecordStatus.LogError)
            {
                uiText.text = "Recording Error: " + status + ", details: " + msg;
            }
            Debug.Log("RecordStatus: " + status + ", details: " + msg);
        };
    }
    public void StartRecorder()
    {
        videoRecorder.StartRecording();
        cameraRecorder =
            Camera.main.gameObject.AddComponent<CameraRecorder>();
        cameraRecorder.Setup(videoRecorder, null);
    }
    public void StopRecorder()
    {
        if (videoRecorder.StopRecording())
        {
            uiText.text = "Recording stop " + videoRecorder.FilePath;
        }
        else
        {
            uiText.text = "Recording failed";
        }
        if (cameraRecorder)
        {
            cameraRecorder.Destroy();
        }
    }
}
```

4．设置脚本

选中 SceneMaster 游戏对象，将 VideoRecorder 游戏对象拖到脚本的 Video Recorder 属性中为其赋值；将 Text 游戏对象拖到 Ui Text 属性中为其赋值，如图 4-13 所示。

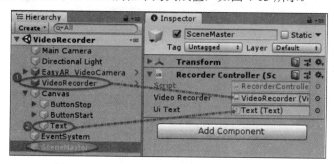

图 4-13

选中 ButtonStart 游戏对象，单击 On Click 标签下的"+"按钮，添加 On Click 事件；将 SceneMaster 游戏对象拖到 On Click 下的框中，设置事件响应的方法来自 SceneMaster 游戏对象；单击右边的下拉按钮，选择 RecorderController 下的 StartRecorder，即设置事件响应的方法是 RecorderController 脚本下的 StartRecorder 方法，如图 4-14 所示。

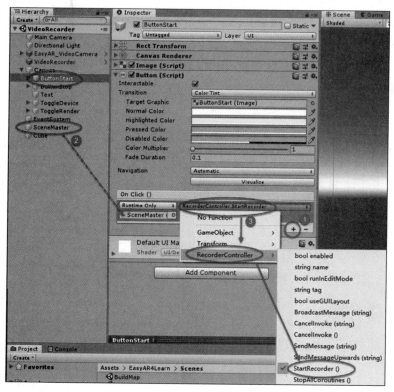

图 4-14

用同样的方法设置"ButtonStop"游戏对象的 On Click 事件响应的方法，把它设置为 SceneMaster 游戏对象下 RecorderController 组件的 StopRecorder 方法。

87

运行以后，单击按钮就可以开始录像，文本框会显示开始录制的提示"Recording start"；单击停止按钮，就会停止录像，如图 4-15 所示。

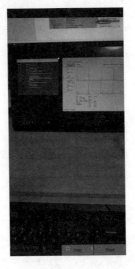

图 4-15

录像的结果会保存在持久数据目录下（具体目录位置请查看 Unity 官方文档 https://docs.unity3d.com/2018.4/Documentation/ScriptReference/Application-persistentDataPath.html），如图 4-16 所示。

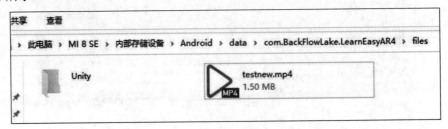

图 4-16

第 5 章
制作涂涂乐和3D跟踪物体例子

5.1 制作涂涂乐

涂涂乐本质是将一个3D模型的UV展开图作为识别图片,然后动态地将识别图片作为贴图贴到模型上。

内容准备

(1)在 Unity 商城里面找一个合适的模型,不要太复杂,关键是贴图只能是一张图片,而且需要有一些留白。这里选取的是如图 5-1 所示的卡通汽车模型。在 Unity 商城中导入资源包,模型样子如图 5-2 所示。

图 5-1

(2)导入资源以后,在 CartoonCarFree/models/Materials 目录下找到模型的贴图,如图 5-3 所示。贴图内容如图 5-4 所示。

(3)用图片编辑器修改贴图,去掉贴图的主要颜色,并把边界线留出来。为了提高贴图的识别率,在空白处添加文字内容,修改结果如图 5-5 所示。

将贴图导入 StreamingAssets 目录下作为识别图片,如图 5-6 所示。

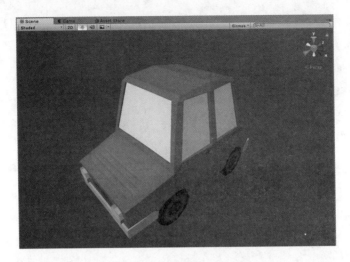

图 5-2

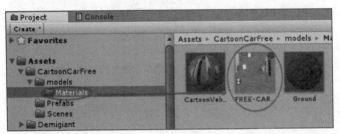

图 5-3

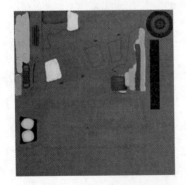

图 5-4

图 5-5

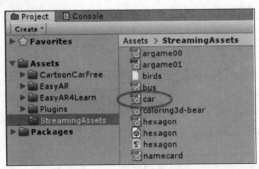

图 5-6

5.2 场 景 制 作

1. 添加基础内容

在 EasyAR 4Learn/Scenes 目录下新建一个场景并命名为 Car-Coloring3D，设置场景中 Main Camera 的 Clear Flags 属性为 Solid Color。

将 EasyAR/Prefabs/Composites 目录下的 EasyAR_ImageTracker-1 预制件拖到场景中，如图 5-7 所示。

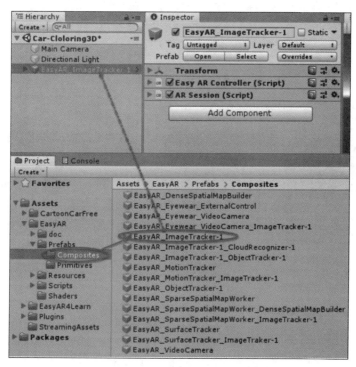

图 5-7

将 EasyAR/Prefabs/Primitives 目录下的 ImageTarget 预制件拖到场景中，设置识别图片是 StreamingAssets 目录下的 car.jpg，如图 5-8 所示。

2. 添加模型

将目录 CartoonCarFree/Prefabs 目录下的 FreeCar 模型拖到 ImageTarget 游戏对象下，设置模型的大小、位置和角度到合适的情况，如图 5-9 所示。

3. 添加涂涂乐内容

选中 FreeCar 游戏对象，将目录 EasyAR 4Learn/Materials 下的 Sample_TextureSample 贴图拖到 Materials 属性下为其赋值，如图 5-10 所示。

将 EasyAR 4Learn/Scripts 目录下的 Coloring3D 脚本拖到 FreeCar 游戏对象下成为其组件，将 RenderCamera 游戏对象拖到 Camera Renderer 属性中为其赋值，如图 5-11 所示。

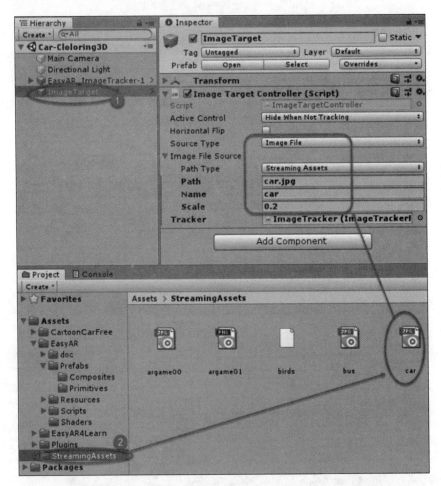

图 5-8

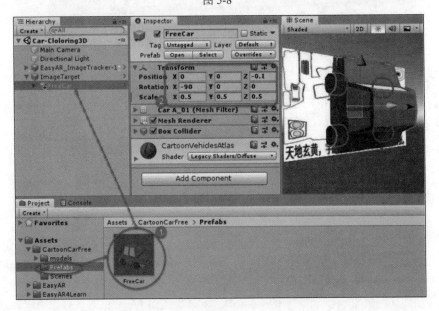

图 5-9

第 5 章 制作涂涂乐和 3D 跟踪物体例子

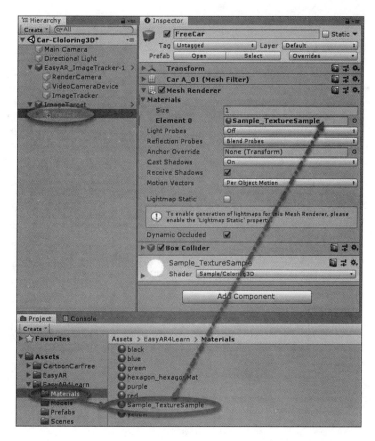

图 5-10

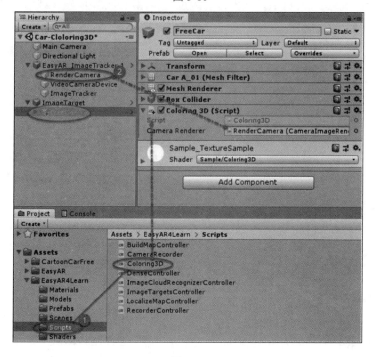

图 5-11

选中 FreeCar 游戏对象下的所有子游戏对象，将 EasyAR 4Learn/Scripts 目录下的 Coloring3D 脚本拖到游戏对象下成为其组件；将 RenderCamera 游戏对象拖到 Camera Renderer 属性中为其赋值，如图 5-12 所示。

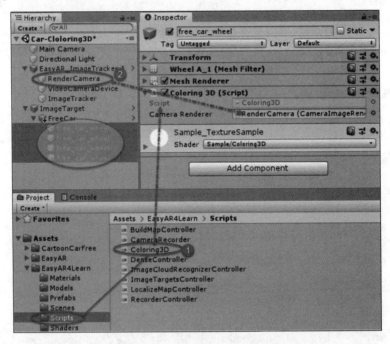

图 5-12

这个时候运行，识别图片以后，在图片对应位置填色或绘制内容，模型对应位置也会一起变化，如图 5-13 所示。

图 5-13

这个例子主要是为了更好地理解涂涂乐的实现过程。因为实际使用的模型与涂涂乐使用的模型 UV 展开的方式不一样，所以如果需要做一个像官方例子那样的涂涂乐，首先需要选择模型并重新展开 UV，重新设计贴图。如何展开 UV 是 3D 建模过程中的内容，就不在这里说明了。

5.3 制作 3D 跟踪物体

EasyAR 识别物体需要一组 obj 文件用于识别，同时还需要有实体的模型。最理想的情况是有专人设计模型，然后用 3D 软件制作并绘制纹理。当然，也可以先用 3D 软件设计，然后利用 3D 打印技术打印出来。

这两种方法都很麻烦，因为这里的目的只是演示，所以这里先在 Unity3D 商城里寻找合适的模型，然后导出成 obj 文件，再通过软件转换成纸模，打印后粘贴制作。

5.3.1 寻找合适的模型

为了节省工作量，需要寻找一个面数比较少的模型，这样不会把大量的时间浪费在模型制作上。另外，模型的纹理要比较丰富，适合识别使用。

这里是在 Unity3D 的商城中寻找模型，当然也有一些软件可以从游戏中导出模型，有兴趣的话读者可以自己尝试一下。

（1）在浏览器中打开 Unity3D 的商城。虽然 Unity3D 也可以打开商城，但是没有在浏览器中操作方便。

（2）单击 3D 选项，选择下面的一个大类，这里选择的是"交通工具"。可以根据自己的兴趣选择其他的，只是不建议选择植物，会很难做，如图 5-14 所示。

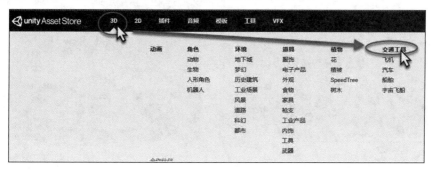

图 5-14

（3）选择"价格"排序，这样可以把免费的排在前面，如图 5-15 所示。

图 5-15

（4）看到合适的模型，移动鼠标到项目上以后单击"快速浏览"按钮，如图 5-16 所示。

图 5-16

（5）很多项目都会给出图片效果，单击左侧的图片列表可以看到具体的效果。图 5-17 所示的消防车感觉比较适合，这里就选它了。

（6）在 Unity3D 编辑器中打开商城界面，将名称输入进去后搜索项目，如图 5-18 所示。

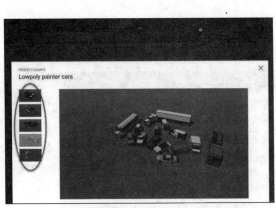

图 5-17　　　　　　　　　　　　　　图 5-18

（7）找到项目以后下载导入项目，如图 5-19 所示。

图 5-19

（8）导入项目以后找到模型，如图 5-20 所示。注意，这里要找到的是 fbx 文件，而不是 Unity3D 的 Prefab 文件。

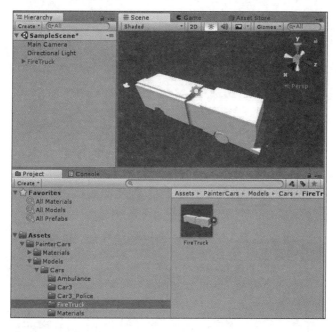

图 5-20

5.3.2 模型修改

这个模型不是单独的，是由 4 个模型组成的，如图 5-21 所示。

（1）在 3D 文件类型转换过程中，有时候会导致模型件的位置关系出错，同时也为了减少工作量，先把零散的部件去掉，只留下最主要的部分，如图 5-22 所示。

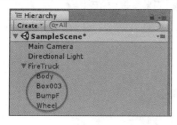
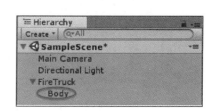

图 5-21　　　　　　　　　　　　　　图 5-22

删除的过程中会提示破坏了原有的 Prefab，单击"Continue"按钮继续就好了，如图 5-23 所示。

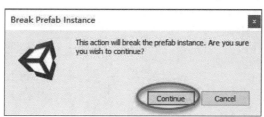

图 5-23

（2）将最主要的模型内容 Body 拖到顶层，修改 Transform 信息为初始值，如图 5-24 所示。

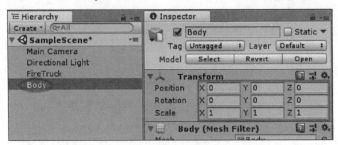

图 5-24

（3）为了方便，将游戏对象的名称由 Body 改为 FireTruckII，并拖到 Project 窗口的 Assets 目录中，成为一个 Prefab，如图 5-25 所示。

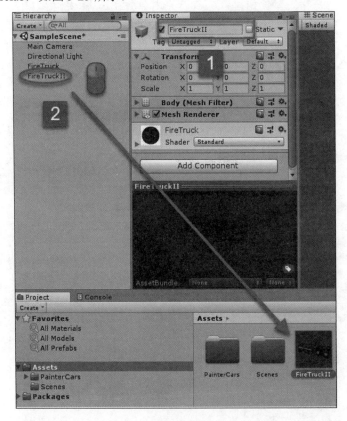

图 5-25

（4）将位于 Assets/PainterCars/Textures/Cars/FireTruck 目录下的 FireTruck.png 文件及该模型的纹理复制一份，改名为 FireTruckII.png，移动到 Assets 的根目录，和刚才生成的 Prefab 文件放在一起，如图 5-26 所示。

（5）这个模型的纹理比较容易识别。如果想要识别得更好，可以自己给纹理添加一些内容，比较简单的方法就是添加一些文字在上面。

模型原本的纹理如图 5-27 所示，修改后如图 5-28 所示。

第 5 章 制作涂涂乐和 3D 跟踪物体例子

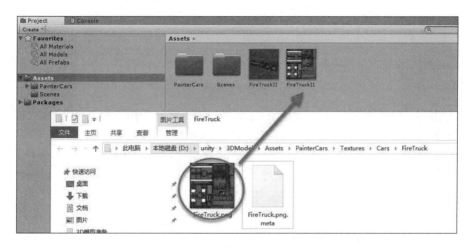

图 5-26

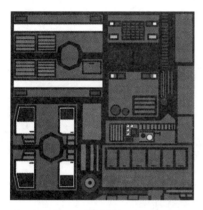

图 5-27

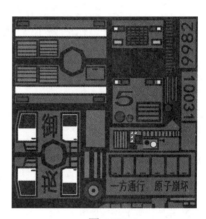

图 5-28

（6）将纹理拖到模型上，看一下效果。

（7）将游戏对象FireTruckII拖到Prefab预制件FireTruckII上，更新保存一次，如图5-29所示。

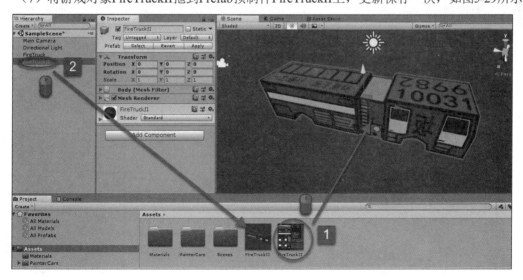

图 5-29

5.3.3 模型导出和转换

模型导出时先通过插件将模型导出成 fbx 格式，然后通过软件转换成 obj 格式。obj 格式既可以给 EasyAR 识别时使用，又可以导入纸模制作软件。

（1）在 Unity3D 的商城中找到 FBX Exporter 插件并导入，如图 5-30 所示。

图 5-30

（2）选中之前生成的 Prefab 文件，选择菜单 GameObject→Export To FBX，如图 5-31 所示。

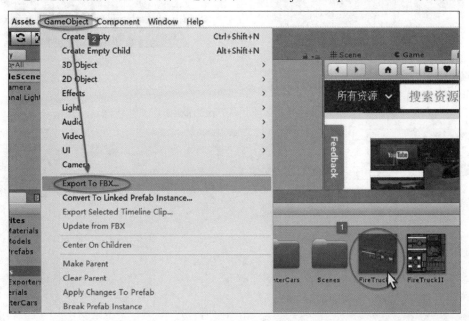

图 5-31

（3）在弹出的窗口中选择导出的路径 Export Path，在 Include 选项中选择只导出模型，不导出动画，单击 Export 按钮即可，如图 5-32 所示。

（4）这时，在导出目录中就会多出一个 FireTruckII.fbx 文件，如图 5-33 所示。

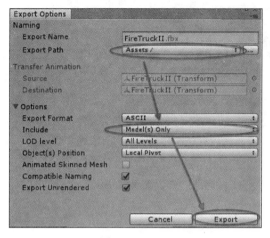

图 5-32

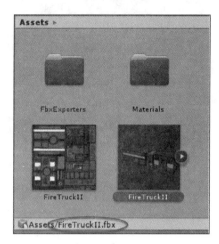

图 5-33

（5）将 fbx 文件和纹理文件复制到一个新的目录中准备转换，如图 5-34 所示。

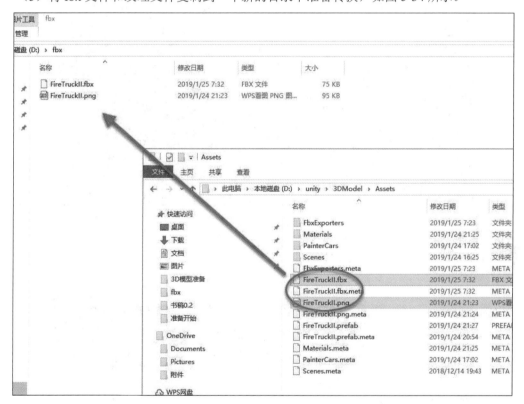

图 5-34

（6）AutoDesk 公司提供了一款免费的 3D 模型文件转换工具，叫 FBX Converter。下载并安装，如图 5-35 所示。

启动以后，界面如图 5-36 所示。

图 5-35

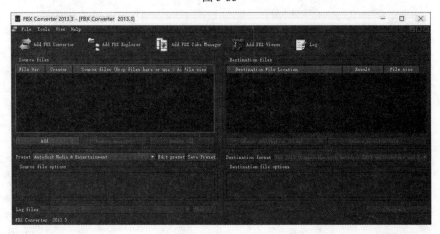

图 5-36

（7）单击左边的 Add 按钮，找到前面复制出来的 fbx 文件，选中并单击"打开"按钮，如图 5-37 所示。

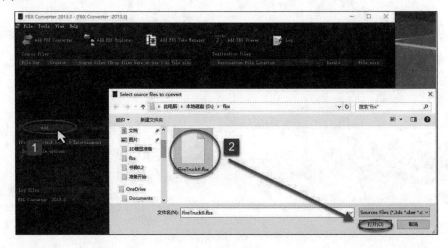

图 5-37

（8）单击右边的 Destination format 旁的下拉按钮，选择导出格式为 OBJ，然后单击 Convert 按钮，即可将 fbx 文件转换成 obj 文件，如图 5-38 所示。

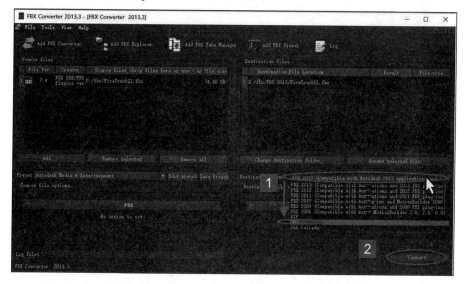

图 5-38

5.3.4 纸模转换制作

纸艺大师是一款纸模设计软件，可以将 3D 模型文件导出成纸模文件，而且它有中文版，使用起来很方便。

（1）下载安装纸艺大师，打开后的界面如图 5-39 所示。

图 5-39

（2）单击打开图标，找到 obj 文件，选中并单击"打开"按钮，如图 5-40 所示。

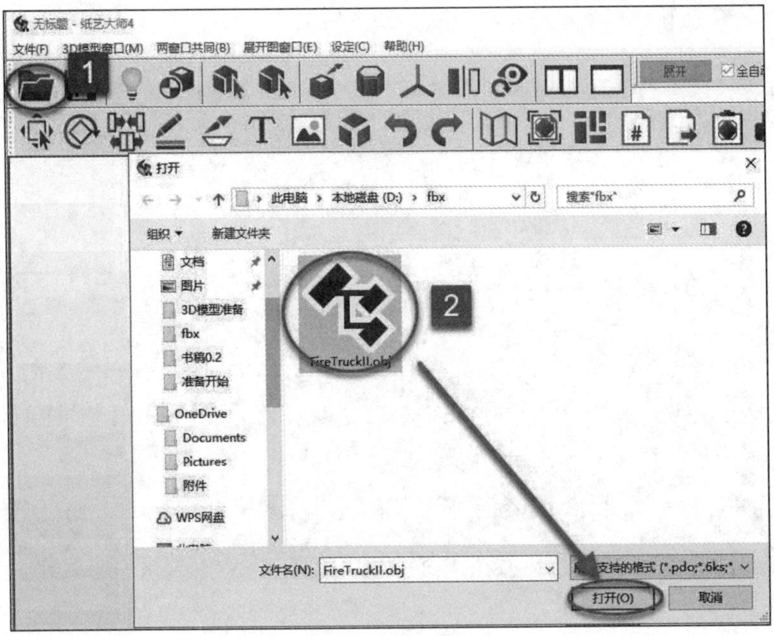

图 5-40

（3）在弹出窗口中单击"关闭"按钮。这里一般不需要更改，除非发现贴图跑到模型内部去了，如图 5-41 所示。

（4）指定大小。这里先单击 OK 按钮（见图 5-42），因为后面可以调整。

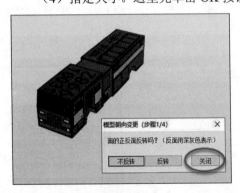

图 5-41

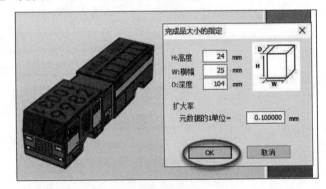

图 5-42

（5）单击菜单上的"展开"按钮，就可以把模型展开，如图 5-43 所示。展开后的效果如图 5-44 所示。

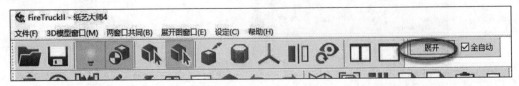

图 5-43

第 5 章 制作涂涂乐和 3D 跟踪物体例子

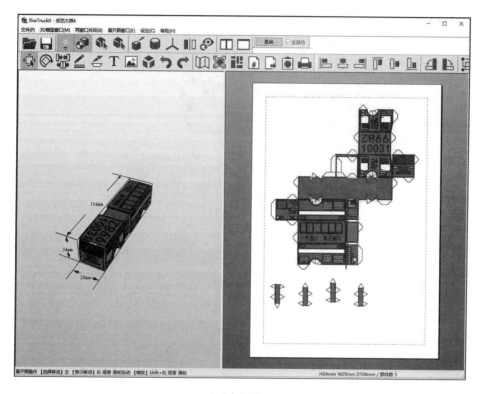

图 5-44

（6）重写调整纸模的粘接位置，太零散或者太大块制作起来都很麻烦。

单击选中菜单上的"面的分离和连接"按钮，如图 5-45 所示。

当鼠标指到右边图片边线上时，出现一条绿色的线，表示可以从绿线处将图片分成两个部分，如图 5-46 所示。

当鼠标指到粘贴部分时，会出现红线和箭头，表示可以从这里将两个部分连接在一起，如图 5-47 所示。

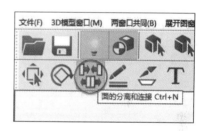

图 5-45

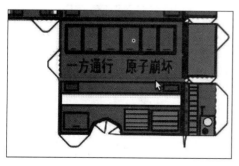

图 5-46

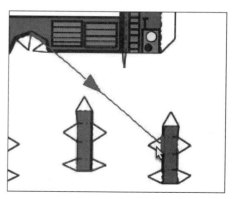

图 5-47

用以上方法调整纸膜粘接位置到合适，如图 5-48 所示。

105

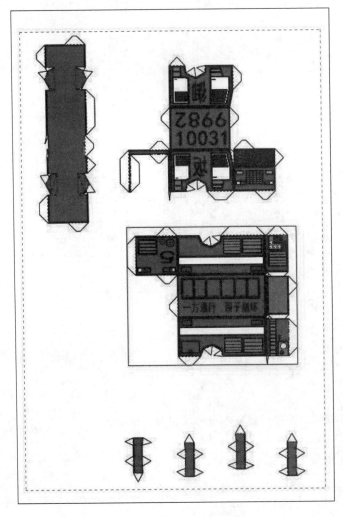

图 5-48

（7）选择菜单中的"设定→变更完成品的大小→指定尺寸和扩大率"，如图 5-49 所示。

（8）修改"高度""横幅"或者"深度"中任一数值即可调整大小，如图 5-50 所示。太大或者太小粘贴起来都很麻烦，但是宁大勿小。

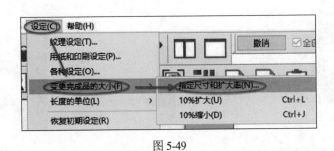

图 5-49

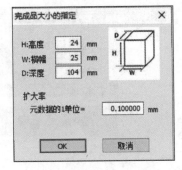

图 5-50

（9）通过修改大小、旋转、移动等方法将纸膜大小调整至合适的尺寸，如图 5-51 所示。

第 5 章 制作涂涂乐和 3D 跟踪物体例子

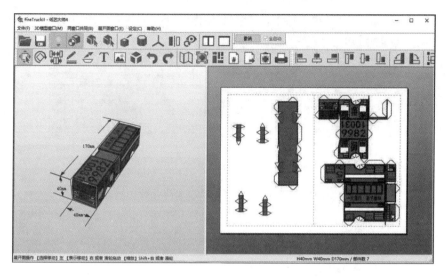

图 5-51

（10）单击菜单"文件→PDF 文件输出"即可将内容导出成 PDF 文件，如图 5-52 所示。

5.3.5 模型制作

将 PDF 文档打印出来，可以是黑白的，剪下来粘好即可。这个用了大约 2 小时，制作过程视频网址为 https://www.bilibili.com/video/av42027324/。

5.3.6 场景制作

1. 导入内容

将模型文件 FireTruckII.obj、firetruckii.mtl 和 FireTruckII.png 导入到 StreamingAssets 目录，作为识别内容，如图 5-53 所示。

将模型文件 FireTruckII.obj、firetruckii.mtl 和 FireTruckII.png 导入到 EasyAR 4Learn/Models 目录，作为识别后显示的内容，如图 5-54 所示。

2. 设置场景

打开 ObjectTarget 场景，单击菜单 File→Save As，将场景另存为 FireTruckII，如图 5-55 所示。

选中 ObjectTarget 游戏对象，设置 Obj Path 属性为 FireTruckII.obj，即要跟踪的.obj 文件的路径；修改 Extra File Paths 的 Size 值为 2，并添加另外 2 个文件的路径，即 FireTruckII.mtl 和 FireTruckII.jpg；设置 Name 属性和 Scale 属性，将跟踪物体改为 FireTruckII，如图 5-56 所示。

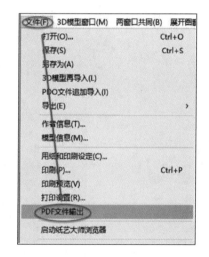

图 5-52

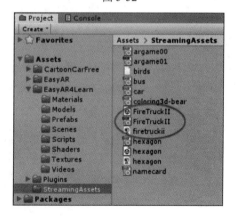

图 5-53

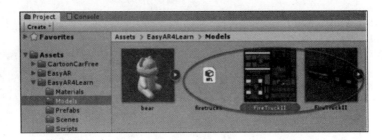

图 5-54

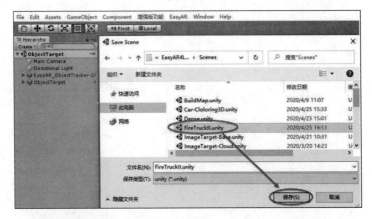

图 5-55

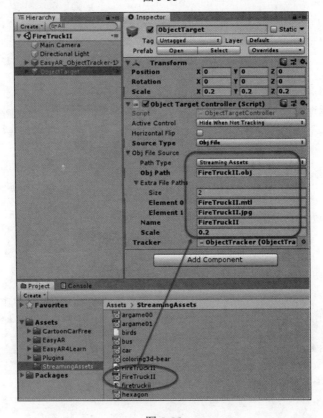

图 5-56

删除 ObjectTarget 游戏对象下原有的子游戏对象；将目录 EasyAR 4Learn/Models 目录下的 FireTruckII 模型拖到 ObjectTarget 游戏对象下成为其子游戏对象。修改 FireTruckII 游戏对象的角度为"90, 0, -180"，如图 5-57 所示。

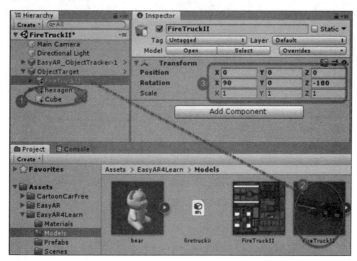

图 5-57

单击运行，就能识别之前做的纸模型了，如图 5-58 所示。

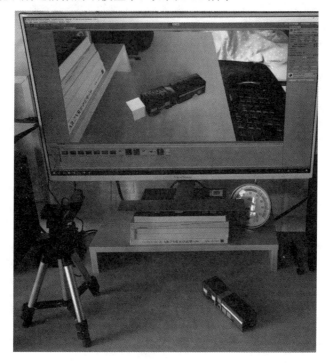

图 5-58

第 6 章 稀疏空间地图室内导航原理

6.1 增强现实室内导航原理说明

6.1.1 基本原理

室内导航与室外导航相比,最大的问题就是难以使用 GPS 来实现定位。室内导航需要的精度更高。另外,在室内,由于建筑物的遮挡,会导致 GPS 和基站信号减弱,无法有效利用室外导航的方式来精确定位。

传统的室内导航通常需要在设施内布设大量的信号点,例如蓝牙、WiFi、可见光等,需要对已有的设施进行改造,并且投入巨大。在增强现实具备了视觉惯性同步定位和建图(VISLAM)技术之后,不需要对原有设施进行改造,仅通过移动设备就能完成室内导航。

在 Unity3D 平台下使用增强现实实现室内导航的原理很简单。例如,面对一个现实空间场景,如图 6-1 所示。

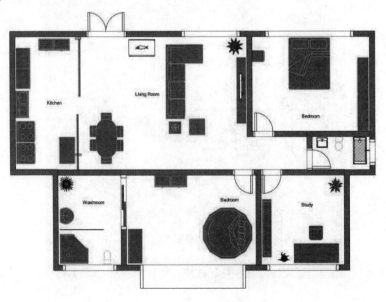

图 6-1

在 Unity3D 的虚拟空间中,建立一个一比一对应的虚拟场景,场景中不需要完全还原现实空间场景的内容,只需要添加导航需要的关键内容,如图 6-2 所示。

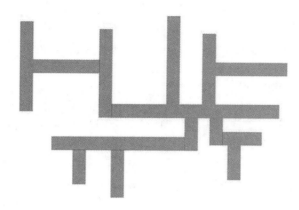

图 6-2

使用导航的时候,将虚拟空间的场景和现实空间的场景进行重合,如图 6-3 所示。

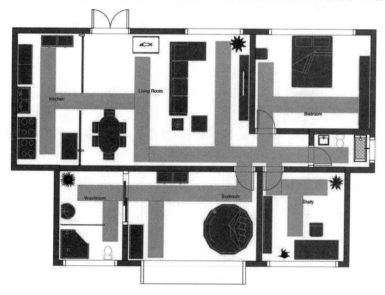

图 6-3

利用 Unity3D 作为游戏引擎自带的导航功能,在虚拟空间场景中进行导航,如图 6-4 所示。

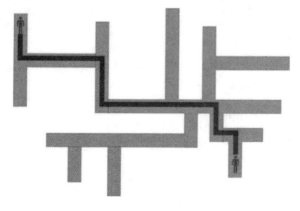

图 6-4

通过增强现实技术让在虚拟空间场景中的导航内容显示出来。因为虚拟空间场景和现实空间场景是重合的，所以显示的导航结果就是现实空间场景中需要的导航结果，如图6-5所示。

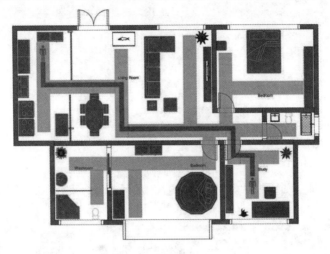

图6-5

具体的实现方式有2种，利用运动跟踪和利用稀疏空间地图。从使用效果而言，建议使用稀疏空间地图。

室内导航的方式不光可以用于室内导航，还可以用于标识室内物体信息，例如标识管道线路走向、设备开关信息等。

6.1.2 利用运动跟踪的实现方式

使用运动跟踪的实现方式，最麻烦的是需要定位当前位置和方向，从而实现现实空间场景和虚拟空间场景的重合。

运动跟踪启动以后，虚拟空间的坐标原点和方向是受启动时设备位置和角度影响的。当运动跟踪功能启动的时候，设备所在位置即是虚拟抗拒的坐标原点，设备背面（设备没有水平放置的时候）是Z轴正方向，如图6-6、图6-7所示。

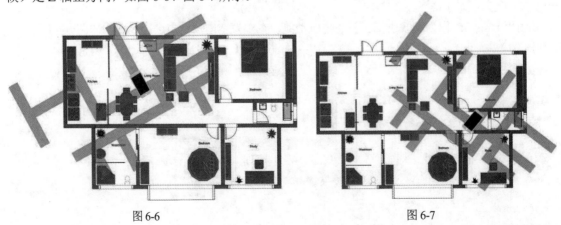

图6-6　　　　　　　　　　　　　图6-7

这个时候，需要想办法让虚拟空间场景和现实空间场景重合，即调整虚拟空间场景的位置和角度，使其和现实空间场景重合，通常有2种方法：利用指南针定位和利用平面图像跟踪定位。

1. 利用指南针定位

利用指南针定位的方式是首先在现实空间设置几个导航起始点,使用者必须站在导航起始点,输入当前起始点的编号或者名称,根据起始点的位置调整虚拟空间场景的位置。利用指南针调整虚拟空间场景的角度,使其和现实空间场景重合。

这种方法的缺点是指南针的精确度和稳定程度会受周围磁场影响,如果附近有强磁场就会干扰指南针的方向导致角度无法正确校正。

2. 利用平面图像跟踪定位

利用平面图像跟踪定位的做法是在现实空间中固定位置贴几个平面图像,使用者必须扫描并跟踪到平面图像以后获取当前平面图像的信息。根据平面图像所在的位置和角度,调整虚拟空间场景的位置和角度,实现现实空间场景和虚拟空间场景的重合。

这种方法的缺点是平面图像跟踪的效果会影响最终导航的效果。

利用运动跟踪的实现方式有以下优缺点。

- 优点:完全不需要与外界交互,或者说可以在完全没有网络的情况下实现室内导航。
- 缺点:首先,只能从指定的位置开始导航。其次,精确度不够,导航空间稍大一些,远端的误差就会被放大。假设开始的时候只有 1 度的角度偏差,虽然这个偏差已经小到肉眼不可见,但是根据三角函数知识可以知道在 10 米外就会产生 0.17 米的偏差。

因此,运动跟踪的实现方式更推荐用于在不大的范围内标识物体信息。

6.1.3 利用稀疏空间地图的实现方式

开始的时候需要在现实空间场景中建立稀疏空间地图,如图 6-8 所示。将虚拟空间场景放置在稀疏空间地图中,如图 6-9 所示。

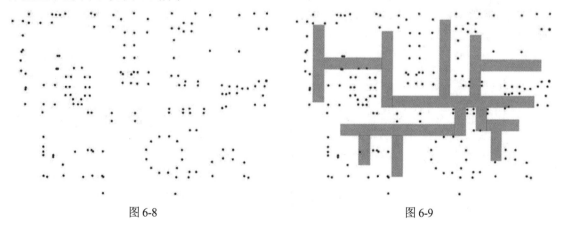

图 6-8   图 6-9

导航开始的时候,利用稀疏空间地图会自动与现实空间场景重合的功能,将虚拟空间场景与现实空间场景重合,从而实现导航功能,如图 6-10 所示。

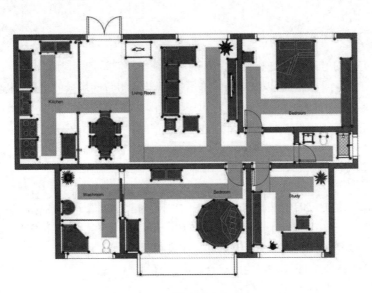

图 6-10

利用稀疏空间地图实现室内导航有以下优缺点。

- 优点：可以从场景中任意一个位置开始导航，没有运动跟踪的角度偏差放大问题。
- 缺点：导航开始时，需要连接网络获取稀疏空间地图。稀疏空间地图建立的质量会影响导航的质量。

6.2 添加虚拟空间场景内容的方式

添加虚拟空间场景内容，如果是运动跟踪的方式，就添加到 WorldRoot 游戏对象下；如果是稀疏空间地图的方式，就添加到 SparseSpatialMap 游戏对象下，如图 6-11 所示。添加的方式有以下 3 种。

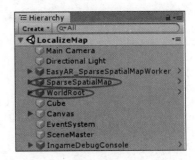

图 6-11

1. 开发时直接添加

如果内容是可以事先确定好的（比如说，有标准的工程图纸，可以按照图纸来建立 3D 模型及导航线路），那么可以在开发的时候直接将虚拟空间场景的内容添加到对应游戏对象下，然后打包编译。

这样做的好处是自由度大，可以添加各种复杂的内容甚至逻辑在里面，但是无法通用。

2. 动态添加手动调整

这种方式是官方示例中的添加方式。应用编译好以后，动态往虚拟空间场景中添加内容，然后通过移动设备界面调整这些内容在空间中的位置。之后将内容在空间的位置和角度信息保存下来，供导航使用。对于一些标识提示性的内容，可以用这样的方式添加，做起来比较直观，通用性也不错。

3. 平面图像跟踪添加

这种方式是上一种方式的补充。因为 EasyAR 4 无法简单准确地将虚拟物体放在同一平面上，所以可以跟踪一个放置在某个平面（一般是地面）上的图像，来获取虚拟空间场景中内容的位置和角度。操作起来比手动调整更方便。

6.3 其 他

稀疏空间地图导航虽然便利，但是使用起来有很多局限。

1. 设备限制

要实现稀疏空间地图导航，必须先支持运动跟踪功能；要实现运动跟踪功能，对设备有硬件要求。简单说，老旧机型和一些低端机型无法支持，具体列表请看官网说明。

2. 环境限制

运动跟踪和稀疏空间地图导航对环境都有一定的要求，需要在有丰富、稳定且静止的视觉特征区域进行，无法在周围有大量移动物体（比如周围人超多）的情况下使用，无法在昏暗的环境下使用。

另外，从技术上而言该技术用于室外导航没有问题，但是实施起来很困难。按照官方说明，每个稀疏空间地图支持 1000 平方米的范围。在室内已经很大，在室外则相对不大。当然，也可以用多个稀疏空间地图来实现，但是应用的复杂程度会提高不少。最麻烦的还是地图的建立。仅靠普通的移动设备，建几十平方米空间的地图，也就两三分钟，但是在室外这个工作量会剧增。除非有专门的设备，否则光建立地图就会花去大量的时间。

第 7 章 项目准备

7.1 总体想法

通过项目演示如何直接添加虚拟内容以及如何动态添加虚拟内容的方法和过程。动态添加虚拟内容意味着需要将添加的内容动态的信息保存下来,此外,还有导航如何实现以及如何动态地生成导航路径等,都是问题。我们先把几个难点解决以后再来设计项目。

通常这样的解决过程并不在项目当中,是另开一个项目解决,完成后就可以删除。考虑到很多读者都是初学者,所以把难点解决的过程也展示出来。

打开 Unity Hub,单击"新建"按钮,输入项目名称为"Kanamori",设置项目所在目录,单击"创建"按钮新建一个项目,如图 7-1 所示。

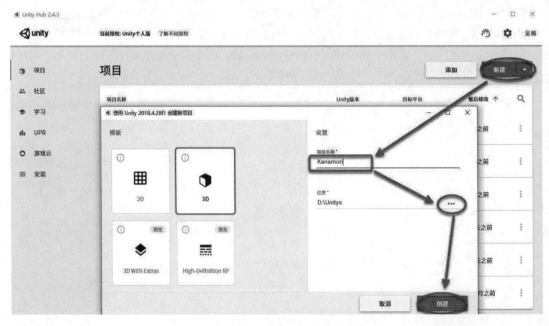

图 7-1

在 Project 窗口中,删除项目中原有的目录 Scenes。单击鼠标右键,在弹出的窗口中单击 Create→Folder 添加目录。在场景中新增 Kanamori 目录,放置所有自己编写添加的内容。在 Kanamori 目录下再添加一个目录 Difficulty,用于放置解决难点的所有内容,如图 7-2 所示。

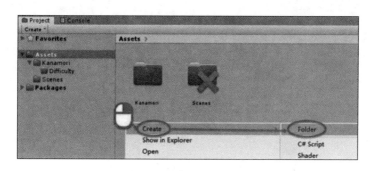

图 7-2

7.2 难点解决

在这里要解决的问题有 3 个：对象信息保存、导航实现和动态生成路径。

7.2.1 对象信息保存

在 Unity3D 当中，保存信息的方式很多，可以用自带的 PlayerPrefs 对象方法，也可以使用数据库。Unity 商城中还有不少专门用于保存信息的插件。综合下来，这里先将对象转换成 JSON 字符串，再将字符串保存到文本文件的方式来实现对象的保存和读取。

1．场景准备

选中 Project 窗口中的 Kanamori/Difficulty 目录，单击鼠标右键，在弹出的菜单中选中 Create→Scene，在 Kanamori/Difficulty 目录中添加一个场景，并将场景命名为 SaveObjects，如图 7-3 所示。

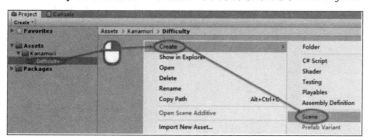

图 7-3

选中 Project 窗口中的 Kanamori/Difficulty 目录，单击鼠标右键，在弹出的菜单中选中 Create→C# Script，在 Kanamori/Difficulty 目录中添加一个脚本，并将脚本命名为 SOController，如图 7-4 所示。

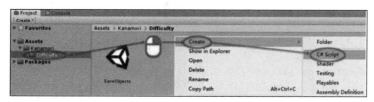

图 7-4

打开 SaveObjects 场景，单击菜单 GameObject→Create Empty，在场景中添加一个空的游戏对象。选中新增的游戏对象，修改其名称为 SceneMaster，并将 SOController 脚本拖到游戏对象上成为其组件，如图 7-5 所示。

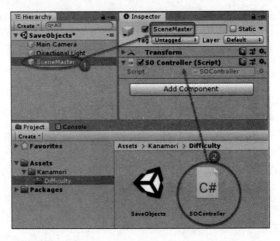

图 7-5

2. 添加基础脚本

因为这里只是为了确定最终使用的方法是否可行，所以尽量减少工作，能使用控制台输出就使用控制台输出。

添加一个命名空间，避免和其他内容冲突：

```
namespace Kanamori.Difficulty
```

添加一个类，命名为 ObjectA。类的属性尽可能和实际用到的接近。在 SOController 类的 Start 方法中添加一个 ObjectA 类的对象并初始化：

```
public class SOController : MonoBehaviour
{
    void Start()
    {
        Debug.Log("start save Object...");

        ObjectA a1 = new ObjectA();
        a1.position = Vector3.zero;
        a1.eulerAngles = new Vector3(45f, 30f, 60f);
        a1.name = "obj-1";
    }
}

public class ObjectA
{
    public Vector3 position;
    public Vector3 eulerAngles;
    public string name;
}
```

3. 添加对象转 JSON 字符串的代码

Unity 脚本自带了相关的转换方法，直接使用即可：

```
string json = JsonUtility.ToJson(a1);
Debug.Log(json);
```

运行场景，在控制台能看到转换后的 JSON 字符串内容，如图 7-6 所示。

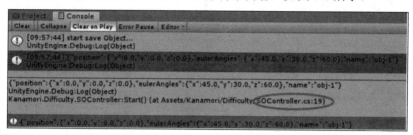

图 7-6

"a1" 对象被转换成以下字符串：

{"position":{"x":0.0,"y":0.0,"z":0.0},"eulerAngles":{"x":45.0,"y":30.0,"z":60.0},"name":"obj-1"}

4. 添加保存字符串的代码

保存字符串到文本文件和读取文本文件的方法在 Unity 官方文档中是没有的，但是可以查微软的 C#文档。

添加引用：

```
using System.IO;
using System;
```

根据文件添加写入脚本：

```
string path = Application.persistentDataPath + "/saveobject.txt";

try
{
   using (StreamWriter writer = new StreamWriter(path))
   {
   writer.WriteLine(json);
   }
   Debug.Log("save end." + path);
}
catch (Exception ex)
{
   Debug.Log(ex.Message);
}
```

运行以后，控制台显示写入情况和位置，如图 7-7 所示。

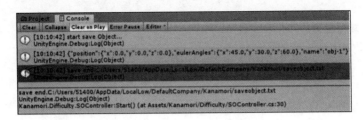

图 7-7

将文件路径复制下来，打开以后就能看到生成的文本文件及其内容，如图 7-8 所示。

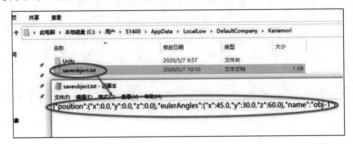

图 7-8

5. 添加读取文本文件脚本

读取功能和写入功能类似：

```
string readJson = "";
try
{
    using (StreamReader reader = new StreamReader(path))
    {
        readJson = reader.ReadLine();
    }
    Debug.Log(readJson);
}
catch (Exception ex)
{
    Debug.Log(ex.Message);
}
```

完成后运行，就能在控制台上看到读取情况，如图 7-9 所示。

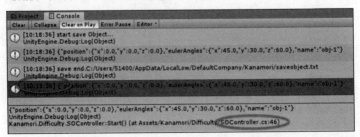

图 7-9

6. 添加字符串转对象脚本

继续利用 Unity 自带的方法进行转换:

```
var objGet = JsonUtility.FromJson<ObjectA>(readJson);
Debug.Log(objGet);
Debug.Log(objGet.name);
```

在控制台上能看到最终效果,如图 7-10 所示。

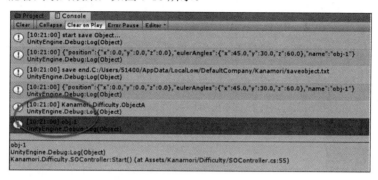

图 7-10

这样就实现了将对象保存到本地文件并从本地文件读取的功能。完整的代码如下:

```
using UnityEngine;
using System.IO;
using System;

namespace Kanamori.Difficulty
{
    public class SOController : MonoBehaviour
    {
        void Start()
        {
            Debug.Log("start save Object...");

            ObjectA a1 = new ObjectA();
            a1.position = Vector3.zero;
            a1.eulerAngles = new Vector3(45f, 30f, 60f);
            a1.name = "obj-1";

            string json = JsonUtility.ToJson(a1);
            Debug.Log(json);

            string path = Application.persistentDataPath + "/saveobject.txt";
            try
            {
                using (StreamWriter writer = new StreamWriter(path))
                {
                    writer.WriteLine(json);
                }
```

```csharp
            Debug.Log("save end." + path);
        }
        catch (Exception ex)
        {
            Debug.Log(ex.Message);
        }
        string readJson = "";
        try
        {
            using (StreamReader reader = new StreamReader(path))
            {
                readJson = reader.ReadLine();
            }
            Debug.Log(readJson);
        }
        catch (Exception ex)
        {
            Debug.Log(ex.Message);
        }
        var objGet = JsonUtility.FromJson<ObjectA>(readJson);
        Debug.Log(objGet);
        Debug.Log(objGet.name);
    }
}
public class ObjectA
{
    public Vector3 position;
    public Vector3 eulerAngles;
    public string name;
}
```

7.2.2 导航实现

导航使用 Unity3D 自带的导航功能，为了使用方便，导入 NavMeshComponents。

1. 获取导航组件

项目地址为 https://github.com/Unity-Technologies/NavMeshComponents。打开项目，单击 Clone or download→Download ZIP 按钮，下载项目，如图 7-11 所示。

下载以后解压，其中是一个完整的 Unity3D 项目及其说明文档，如图 7-12 所示。实际使用到的只是项目的 NavMeshComponents 目录。将该目录拖到自己的项目目录下实现导入，如图 7-13 所示。

第 7 章 项 目 准 备

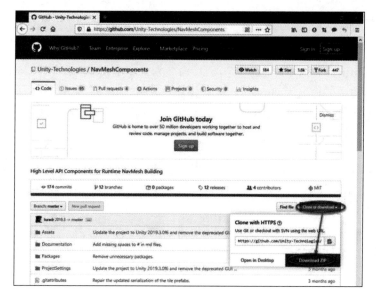

图 7-11

图 7-12

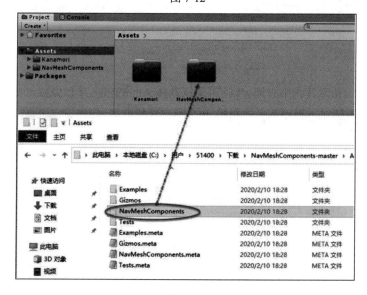

图 7-13

2. 设置基本场景

（1）添加地面

在Kanamori/Difficulty目录下新建一个场景，命名为NavMesh0。单击菜单GameObject→Empty，添加一个空的游戏对象并命名为Ground，用于放置要导航的地面，如图7-14所示。

在Kanamori目录下新建一个目录Textures，添加一些纯色的图像，作为场景中物体的纹理以便区分，如图7-15所示。

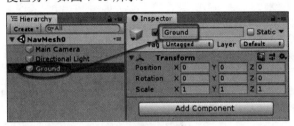 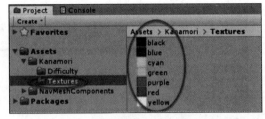

图7-14　　　　　　　　　　　　　　　图7-15

选中Ground游戏对象，单击鼠标右键，在弹出的菜单中选择3D Object→Plane，在Ground游戏对象下添加平面作为地面，如图7-16所示。

修改Plane游戏对象的位置和缩放，使其成为环形；将颜色纹理拖到其上以便区分，这里设置为黄色，如图7-17所示。

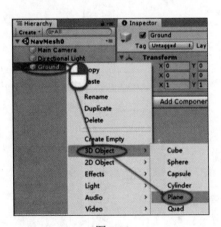 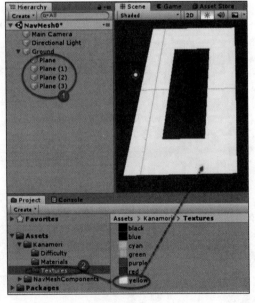

图7-16　　　　　　　　　　　　　　　图7-17

平面游戏对象的位置和缩放参数如表7-1所示。

表7-1　平面游戏对象的位置和缩放参数

位　　置	缩放参数
0, 0, 0	0.3, 1, 1
7, 0, 0,	0.3, 1, 1
3.5, 0, 6.5	1, 1, 0.3
3.5, 0, -6.5	1, 1, 0.3

（2）添加其他游戏对象

单击菜单 GameObject→3D Object→Sphere，在场景中添加一个球体，修改球体游戏对象名称为 TargetBall，用于标识目的地；将纹理拖到球体上，设置球体颜色，这里设置为红色，如图 7-18 所示。

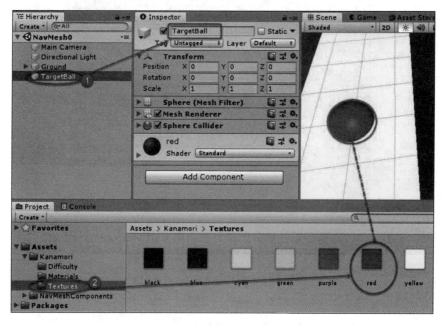

图 7-18

单击菜单 GameObject→3D Object→Capsule，在场景中添加一个胶囊，修改胶囊游戏对象名称为 Player，用于标识玩家或 NPC；将纹理拖到球体上，设置球体颜色，这里设置为蓝色，如图 7-19 所示。

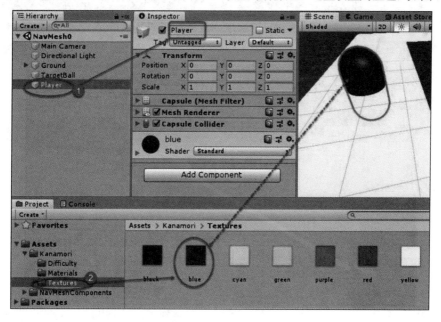

图 7-19

单击菜单 GameObject→Empty，添加一个空的游戏对象并命名为 SceneMaster，用于放置脚本；在 Kanamori/Difficulty 目录中添加一个脚本，并将脚本命名为 NavController，拖到 SceneMaster 游戏对象下成为其组件，如图 7-20 所示。

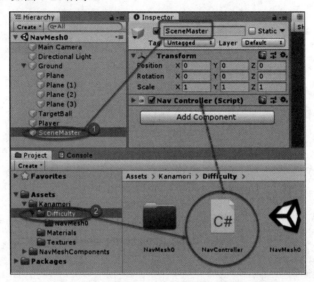

图 7-20

修改 Main Camera 游戏对象的位置和角度，使其能够完整看到地面，即 Ground 游戏对象下的所有内容。这里设置 Position 为 "-6, 6.5, 0"、Rotation 为 "30, 90, 0"，如图 7-21 所示。

图 7-21

3. 基础导航

（1）添加导航代理

选中 Player 游戏对象，单击 Add Component 按钮，在搜索框中输入 "agent"，选中 Nav Mesh Agent，为 Player 游戏对象添加一个导航代理组件，如图 7-22 所示。

（2）添加导航面

选中 Ground 游戏对象，单击 Add Component 按钮，选中 Navigation→NavMeshSurface，为其添加一个导航面组件，如图 7-23 所示。

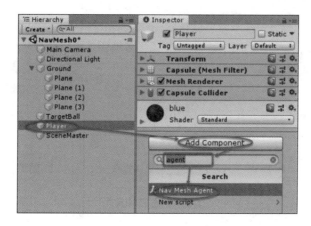

图 7-22

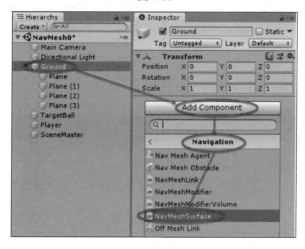

图 7-23

设置 Collect Objects 属性为 Children，这样导航面只会影响 Ground 游戏对象的子对象，而不会影响其他游戏对象；单击 Bake 按钮，烘焙导航面，如图 7-24 所示。

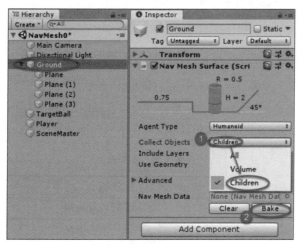

图 7-24

烘焙完成后，会在 Difficulty 目录下新增一个目录，名为 NavMesh0，下面放置烘焙后的导航面信息，并且在 Scene 窗口中会用淡蓝色区块显示导航范围，如图 7-25 所示。

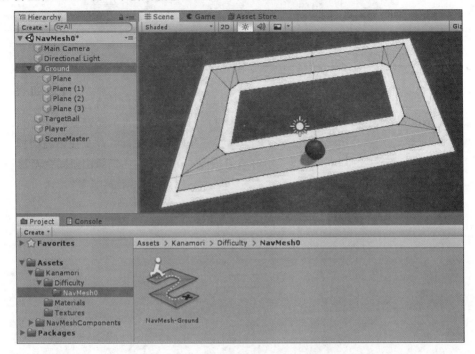

图 7-25

（3）编写设置脚本

在 Update 方法中，用导航代理的 SetDestination 方法就能实现从玩家到目标的移动。

```
public NavMeshAgent agent;
public Transform target;

void Update()
{
    agent.SetDestination(target.position);
}
```

选中 SceneMaster 游戏对象，将 TargetBall 游戏对象拖到 Target 属性中为其赋值，将 Player 属性拖到 Agent 游戏对象中为其赋值，如图 7-26 所示。

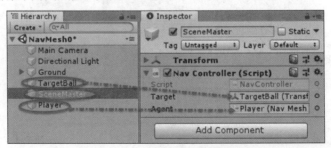

图 7-26

运行结果，Player 游戏对象会自动跑到 TargetBall 游戏对象位置，但不会离开导航面。在 Scene 场景中拖动 TargetBall 游戏对象，就能看到 Player 游戏对象追着跑的情形，如图 7-27 所示。

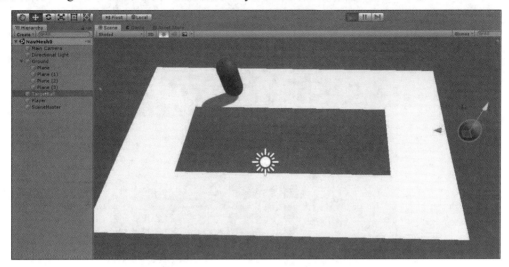

图 7-27

4．显示导航线

实际使用中，不需要自动运动到目标，但是需要显示到目标点的路径。

（1）添加线条特效

单击菜单 GameObject→Create Empty，在场景中添加一个空的游戏对象，修改游戏对象名称为 Line；单击 Add Component→Effects→Line Renderer，为其添加一个线条特效的组件，如图 7-28 所示。

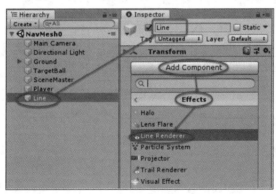

图 7-28

（2）设置线条材质

单击"Materials"属性下的元素，设置其材质为 Sprites-Default，如图 7-29 所示。

（3）设置线条宽度

修改 Positions 属性，设置其 Size 值大于 2，修改元素中各个点的位置，使其能在 Scene 窗口中看得比较明显；拖动 Width 属性下的线条，设置线条宽度，如图 7-30 所示。Width 属性可以设置为一条线的各个位置宽度不一样。

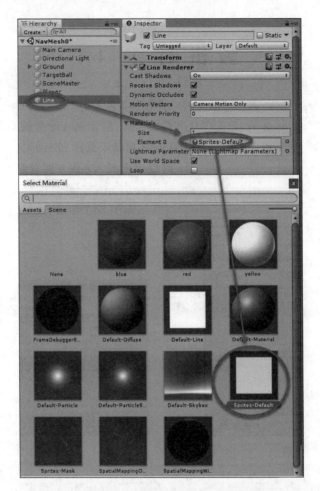

图 7-29

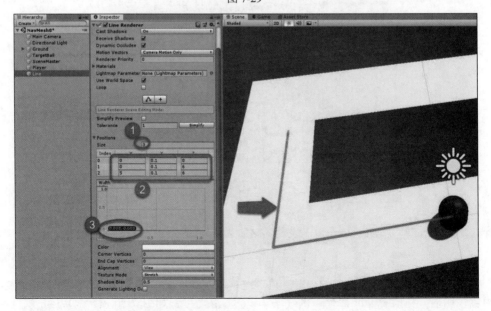

图 7-30

这里不需要那么复杂，整条线的宽度一致，看后面那个数值即可，单位是米，如图 7-31 所示。

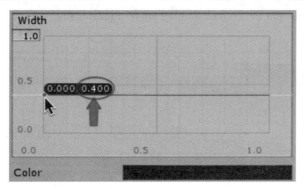

图 7-31

（4）设置线条颜色

单击 Color 属性，在 Gradient Editor 弹出窗口中设置颜色为黑色，如图 7-32 所示。线条颜色也可以设置成多个，这里设置整条线都是黑色的。

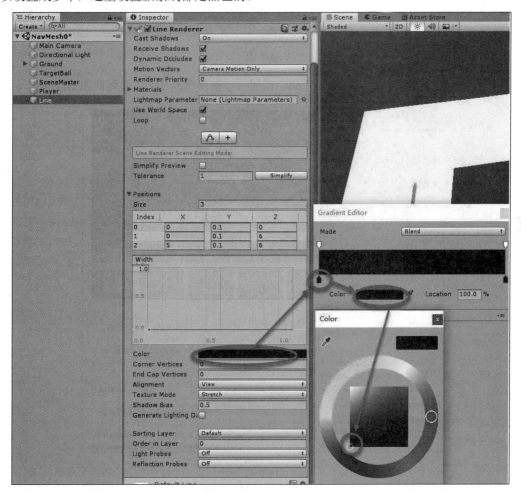

图 7-32

（5）修改并设置脚本

用导航代理的 CalculatePath 方法就能计算出到目的地的路径。设置线条特效的 positions 属性为路径即可。

```
public LineRenderer lineRenderer;
private NavMeshPath path;

void Start()
{
    path = new NavMeshPath();
}

void Update()
{
    //agent.SetDestination(target.position);
    agent.CalculatePath(target.position,path);
    lineRenderer.positionCount = path.corners.Length;
    lineRenderer.SetPositions(path.corners);
}
```

选中 SceneMaster 游戏对象，将 Line 游戏对象拖到 Line Renderer 属性中为其赋值，如图 7-33 所示。

图 7-33

在 Scene 窗口中拖动 TargetBall 游戏对象，就能看到一条黑线连接 TargetBall 游戏对象和 Player 游戏对象，也就是导航路径。此时，选中 Line 游戏对象，能看到具体的点和点的位置，如图 7-34 所示。

选中 Ground 游戏对象，就会显示出导航面。这时能看到导航路径是选取的最短路径，即会沿着导航面的边进行导航，如图 7-35 所示。

此时拖动 Player 游戏对象会发现 Player 游戏对象必须在导航面范围内，即导航代理必须在导航面范围内。如果启动时导航代理在导航面外不远的地方，则启动以后 Player 游戏对象会自动跳到最近的导航面上。如果启动时导航代理在导航面外很远的地方，则会出现错误提示 ""CalculatePolygonPath" can only be called on an active agent that has been placed on a NavMesh."。

第 7 章 项目准备

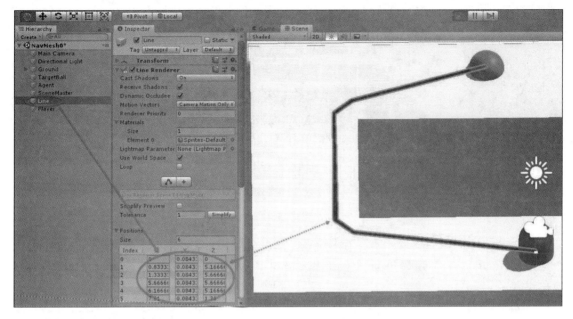

图 7-34

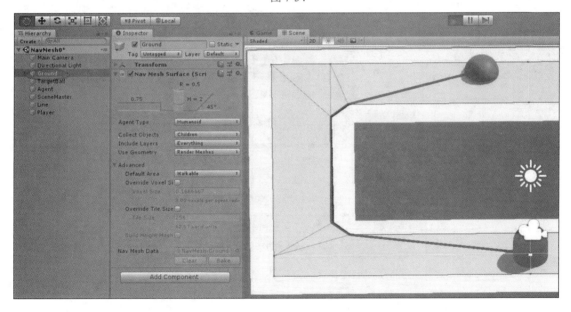

图 7-35

5. 导航线居中

导航线如果太靠边，稍微偏差导航线就到墙里去了，看上去的效果会贴着墙。为了看上去的效果是导航线在中间，可以通过把导航的路径变窄，然后使得导航面变成很窄的细条来实现。

（1）复制原有路径

复制 Ground 游戏对象，删除 Ground 游戏对象上的 Nav Mesh Surface 组件；将 Ground 游戏对象下的物体设置成其他颜色，这里设置为绿色，如图 7-36 所示。

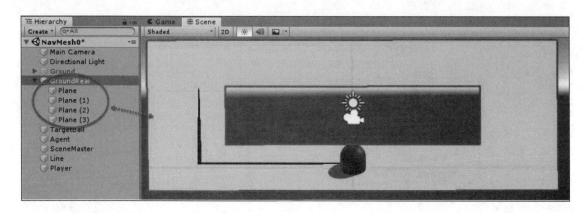

图 7-36

（2）修改原有路径

修改 Ground 游戏对象在 Plane 下，如图 7-37 所示。平面游戏对象的位置和缩放参数如表 7-2 所示。

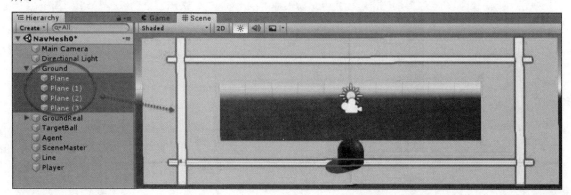

图 7-37

表 7-2 平面游戏对象的位置和缩放参数

位　　置	缩放参数
0，0，0	0.03，1，1.4
7，0，0，	0.03，1，1.4
3.5，0，6.5	1，1，0.03
3.5，0，-6.5	1，1，0.03

（3）修改代理，重新烘焙

单击菜单 Windows→AI→Navigation，打开导航设置窗口；将 Agents 标签下的 Radius 属性修改为 0.1，如图 7-38 所示。

（4）重新烘焙

单击 Ground 游戏对象下的 Bake 按钮，重新烘焙导航面，如图 7-39 所示。

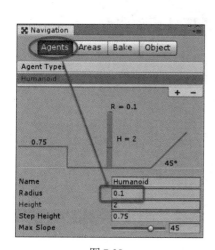

图 7-38

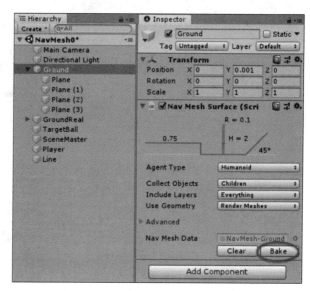

图 7-39

此时再运行，导航线基本就在路的中间了，如图 7-40 所示。

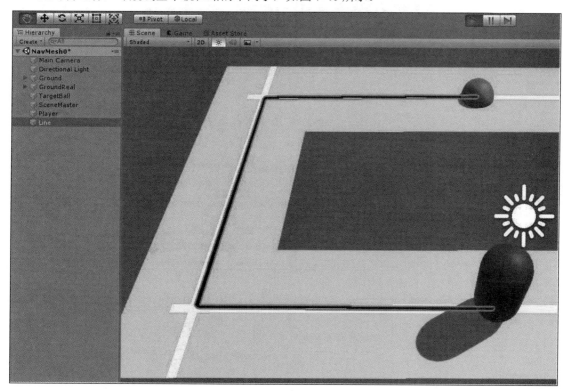

图 7-40

6. 玩家在导航面之外导航

现在这样设置的导航面很小，使用者一定会出现在导航面之外。为了让使用者在路两边也能导航，需要把导航代理和玩家分离，可以利用导航代理会自己跑到最近的导航面的效果来实现。

选中 Player 游戏对象，单击导航代理组件右边的按钮，在弹出的菜单中选择 Remove Component 删除组件，如图 7-41 所示。

图 7-41

新建一个空的游戏对象，并为其添加导航代理，如图 7-42 所示。

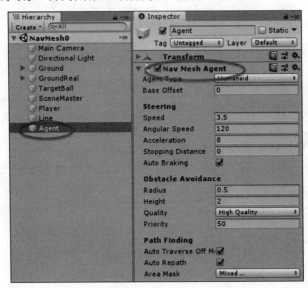

图 7-42

修改脚本，添加一个玩家属性，在每次导航线显示之前停用导航代理，将其位置移动到玩家位置，再启用导航代理。

```
public Transform player;

void Update()
{
    agent.enabled=false;
    agent.transform.position=player.position;
    agent.enabled=true;
    ...
}
```

选中 SceneMaster 游戏对象，将 Agent 游戏对象拖到 Agent 属性中为其赋值，将 Player 游戏对象拖到 Player 属性中为其赋值，如图 7-43 所示。

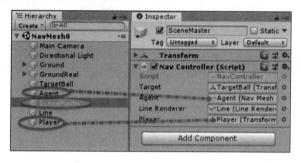

图 7-43

此时运行，Player 游戏对象就能被拖动到导航面之外，仍然能显示导航线，如图 7-44 所示。

图 7-44

导航完整的代码如下：

```
using UnityEngine;
using UnityEngine.AI;

namespace Kanamori.Difficulty
{
    public class NavController : MonoBehaviour
    {
        public Transform target;
        public NavMeshAgent agent;
        public LineRenderer lineRenderer;
        private NavMeshPath path;
        public Transform player;
```

```
void Start()
{
    path = new NavMeshPath();
}
void Update()
{
    //agent.SetDestination(target.position);
    agent.enabled=false;
    agent.transform.position=player.position;
    agent.enabled=true;
    agent.CalculatePath(target.position,path);
    lineRenderer.positionCount = path.corners.Length;
    lineRenderer.SetPositions(path.corners);
}
}
}
```

7．动态生成路径

动态生成路径的关键点是如何用一种简单的方法将虚拟空间中不同的点用 3D 模型连接起来。它的基本实现思路是，将一个平面放置到 2 个点的中间，修改其角度和长度，利用平面将 2 个点连接在一起。

（1）添加基本内容

选中 Project 窗口中的 Kanamori/Difficulty 目录，单击鼠标右键，在弹出的菜单中选中 Create→Scene，在 Kanamori/Difficulty 目录中添加一个场景，并将场景命名为 Paths。

打开场景，单击菜单 GameObject→3D Object→Sphere，在场景中添加一个球体，并命名为 PointA。修改球体颜色为黄色。

复制 PointA 游戏对象为 PointB。

单击菜单 GameObject→3D Object→Plane，在场景中添加一个平面，如图 7-45 所示。

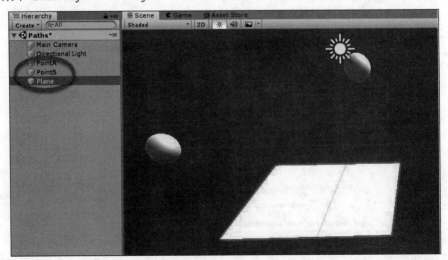

图 7-45

在 Kanamori/Difficulty 目录中添加一个脚本，并将脚本命名为 PathsController。

单击菜单 GameObject→Create Empty，在场景中添加一个空的游戏对象。选中新增的游戏对象，修改其名称为 SceneMaster，并将 PathsController 脚本拖到游戏对象上成为其组件，如图 7-46 所示。

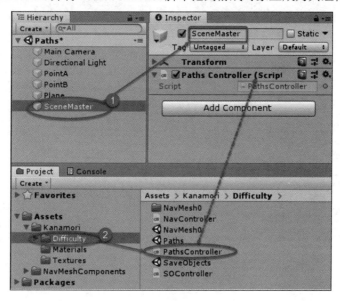

图 7-46

（2）将平面放置到 2 个点中间

修改脚本，在 Update 事件中将平面放置到 2 个点的中间。

```
public Transform pa;
public Transform pb;
public Transform path;

void Update()
{
    path.position = (pa.position + pb.position) / 2;
}
```

选中 SceneMaster 游戏对象，将 PointA 游戏对象拖到 Pa 属性中为其赋值，将 PointB 游戏对象拖到 Pb 属性中为其赋值，将 Plane 游戏对象拖到 Path 属性中为其赋值，如图 7-47 所示。

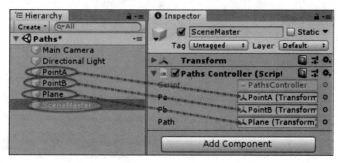

图 7-47

运行效果如图 7-48 所示。可以随意拖动 PointA 或 PointB 游戏对象，Plane 游戏对象始终在它们中间。

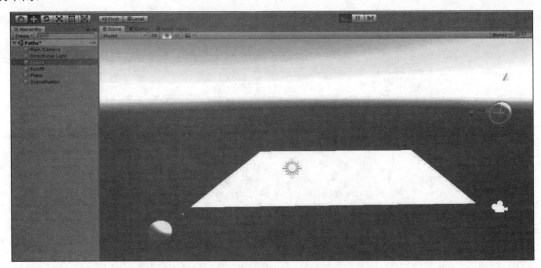

图 7-48

（3）设置角度

角度可以通过两点之间的矢量与 X、Y、Z 平面投影计算出具体的角度。Unity3D 中有一个 LookAt 方法，可以令某个 Transform 始终面向一个点。利用这个方法调整角度就很容易实现。

添加 LookAt 方法，看向 Pa、Pb 都可以。

```
void Update()
{
    path.position = (pa.position + pb.position) / 2;
    path.LookAt(pa);
}
```

此时运行，Plane 游戏对象能自动调整角度到合适的位置，如图 7-49 所示。

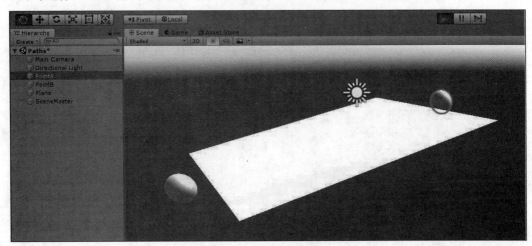

图 7-49

修改Plane游戏对象的Scale属性为"0.03,1,1",平面就会变成一条细的路径,如图7-50所示。

图 7-50

（4）修改长度

修改长度就是对路径的 localScale.z 属性进行修改,使其等于两点之间的距离。

修改脚本:

```
void Update()
{
    path.position = (pa.position + pb.position) / 2;
    path.LookAt(pa);
    path.localScale = new Vector3(0.03f, 1f,
            (pa.position - pb.position).magnitude * 0.1f);
}
```

此时运行,随意拖动PointA或PointB游戏对象,Plane游戏对象始终能将2个点连接在一起,如图 7-51 所示。

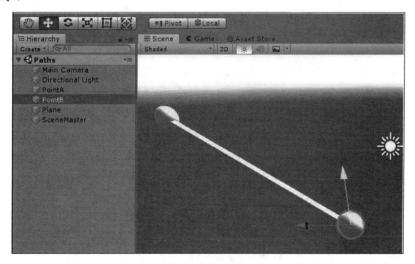

图 7-51

完整脚本内容如下：

```
using UnityEngine;

namespace Kanamori.Difficulty
{
    public class PathsController : MonoBehaviour
    {
        public Transform pa;
        public Transform pb;
        public Transform path;

        void Update()
        {
            path.position = (pa.position + pb.position) / 2;
            path.LookAt(pa);
            path.localScale = new Vector3(0.03f, 1f,
              (pa.position - pb.position).magnitude * 0.1f);
        }
    }
}
```

7.3 项目设计

7.3.1 场景设计

应用运行的主要流程包括 3 部分，即新建地图、添加虚拟内容、实现导航，如图 7-52 所示。

图 7-52

其中，往地图中添加虚拟内容的方法有 3 种（开发时直接添加、动态添加动调整、平面图像跟踪添加），希望 3 种方式都能展现。实现导航的具体步骤包括地图本地化、添加虚拟内容、实现导航，如图 7-53 所示。

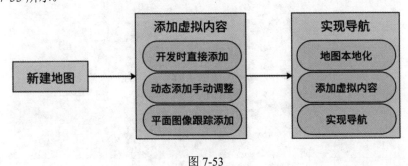

图 7-53

添加虚拟内容的部分可以分成 2 个场景实现，这样每个场景都会比较简单。实现导航的内容只能在一个场景中实现，无法拆分。为了能在各场景中跳转，还需要一个菜单场景，如图 7-54 所示。

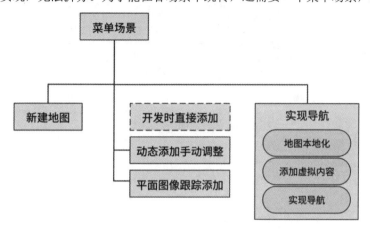

图 7-54

其中，"平面图像跟踪添加"方式用于添加导航路径上的关键点，"动态添加手动调整"方式用于添加其他模型，"开发时直接添加"方式用于添加其他模型。

整个应用至少需要 5 个场景：菜单、地图、模型（动态添加手动调整）、关键点（平面图像跟踪添加）、导航。

添加的关键点还需要设置相互的关系，即路径是产生在哪 2 个关键点中间，所以还需要增加一个"路径"场景，用于设置地图。场景说明如表 7-3 所示。

表 7-3　场景说明

使用场景	名　　称	说　　明
菜单	Menu	跳转菜单
地图	Map	建立稀疏空间地图
模型	Model	动态添加设置模型
关键点	KeyPoint	平面图像跟踪添加设置关键点
路径	Road	添加设置路径
导航	Navigation	实现导航

因为 EasyAR 4 的稀疏空间地图无法在 Unity3D 的编辑器中运行调试，必须发布到设备上才能调试，很麻烦，所以新建一组在编辑器中运行调试的场景，将与稀疏空间地图无关的内容开发完成后再开发实际使用的场景。这样节省时间，并方便调试和修改。场景命名如表 7-4 所示。

表 7-4　场景命名

使用场景	名　　称	方便调试用场景
菜单	Menu	DbgMenu
地图	Map	DbgMap
模型	Model	DbgModel

（续表）

使用场景	名称	方便调试用场景
关键点	KeyPoint	DbgKeyPoint
路径	Road	DbgRoad
导航	Navigation	DbgNavigation

7.3.2 界面设计

菜单场景界面比较简单，只是把对稀疏空间地图的删除操作和名称设置放了过来。这样在添加地图的时候操作更简单，如图 7-55 所示。

新建稀疏空间地图场景内容很少，主要是保存和相关提示，如图 7-56 所示。

图 7-55

图 7-56

模型场景界面比较复杂，为了能精确操作，有很多按钮，如图 7-57 所示。

关键点场景扫描图像后，显示模型，点击模型再显示编辑界面，如图 7-58 所示。

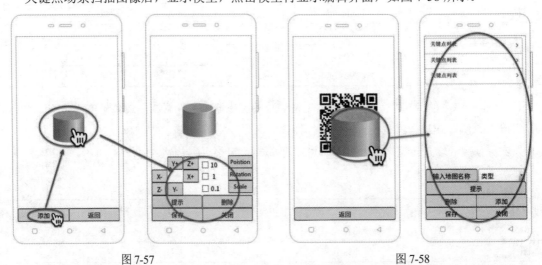

图 7-57　　　　　　　　　　　　图 7-58

路径场景界面和关键点场景编辑界面差不多，如图 7-59 所示。

导航场景在地图定位以后显示目的地列表，点击后，返回上一个界面，如图 7-60 所示。

图 7-59

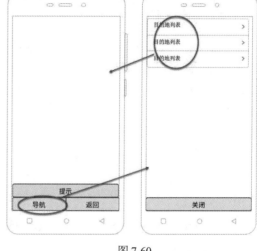

图 7-60

7.3.3 开发模式

项目主要采用 Empty-GameObject 开发模式。每个场景都有一个空的游戏对象，名为 SceneMaster，上面有"场景名+Controller"的脚本。场景的主要逻辑内容都在该脚本中。

整个应用中有一个 GameMaster 游戏对象，此游戏对象不会因为场景切换被删除。上面的 GameController 脚本有 2 个作用：第一，在各场景直接传递信息；第二，放置一些各个场景都会用到的功能。

开发模式的整体结构如图 7-61 所示。

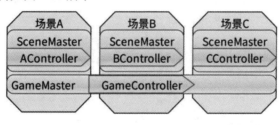

图 7-61

7.3.4 其他内容

- 项目名称：Kanamori
- Unity3D 版本：2018.4.X
- EasyAR 版本：4.0
- 测试硬件：小米 8SE（Android 10）
- APK 版本：Android 8.0
- 屏幕方向：Portrait
- 版本管理：GitHub

7.4 项目搭建

项目中已经有一部分内容了，接下来继续导入一些会用到的内容。

在 Kanamori 目录下新建 Scenes 目录，用于放置开发的场景；在 Scenes 目录下新建 Dbg 目录，用于放置调试场景；在 Kanamori 目录下新建 Scripts 目录，用于放置开发的脚本；在 Scripts 目录下新建 Dbg 目录，用于放置调试脚本，如图 7-62 所示。

在 Kanamori 目录下新建 Fonts 目录，用于放置中文字体，并且导入一个中文字体。

下载并导入 EasyAR 4.0，如图 7-63 所示。

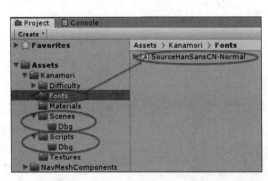
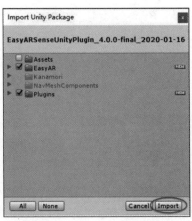

图 7-62　　　　　　　　　　图 7-63

单击菜单 EasyAR→Change License Key，输入 EasyAR 的各种 Key，如图 7-64 所示。

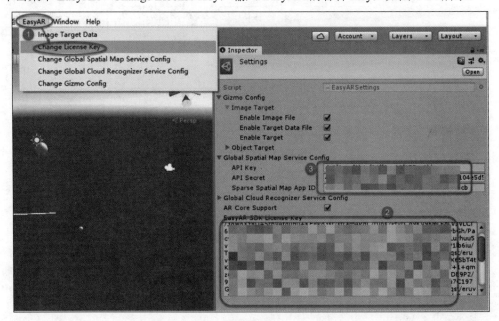

图 7-64

第 7 章 项目准备

单击菜单 File→Build Settings，在 Build Settings 窗口中选中安卓平台，单击 Player Settings 按钮。选中 Other Settings 标签，修改 Package Name 的值与 EasyAR 的 key 对应，修改编译后的安卓版本，如图 7-65 所示。

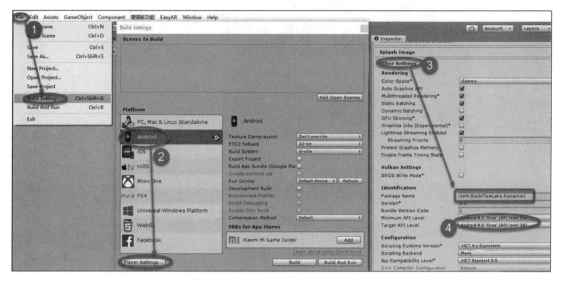

图 7-65

从 Unity3D 的商城导入一个简单的模型，如图 7-66 所示。

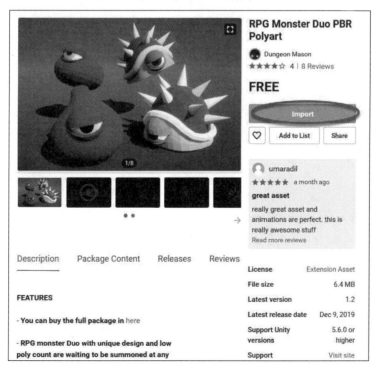

图 7-66

导入 In-game Debug Console 插件，方便调试，如图 7-67 所示。

147

图 7-67

第 8 章 调试场景开发

8.1 菜单场景开发

8.1.1 场景设置

（1）新建场景

在 Kanamori/Scenes/Dbg 目录下新建场景并命名为 DbgMenu。打开 DbgMenu 场景，在场景中添加一个空的游戏对象并命名为 SceneMaster。

在 Kanamori/Scripts 目录下新建脚本并命名为 MenuController。因为菜单场景不涉及稀疏空间地图内容，所以调试场景用的脚本和实际场景的脚本是同一个。

选中 SceneMaster 游戏对象，将 MenuController 拖到其上并设置为其组件，如图 8-1 所示。

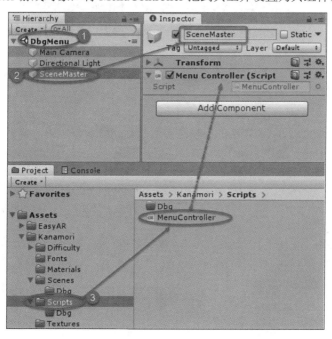

图 8-1

（2）添加设置按钮

单击菜单 GameObject→UI→Button，为场景添加一个按钮，如图 8-2 所示。

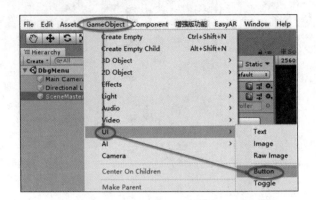

图 8-2

选中按钮，单击 Anchor Presets 按钮，设置对齐方式为底部对齐，如图 8-3 所示。

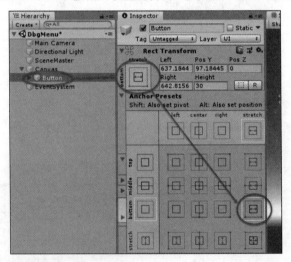

图 8-3

选中移动工具，拖动 Scene 窗口中的小三角向右移动，使其到屏幕宽度的一半，和修改 Anchors 属性 Min X 为 0.5 是一样的效果，如图 8-4 所示。

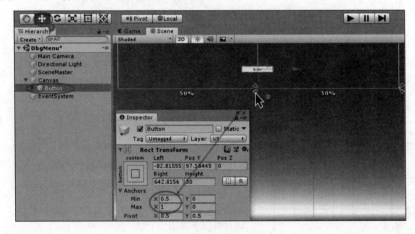

图 8-4

修改按钮的 Rect Transform 属性，"Left，Right，Pos Y，Height，Pos Z"为"0，0，70，140，0"，这样按钮就变成在屏幕右下角、宽度为半个屏幕、高度为 140，如图 8-5 所示。

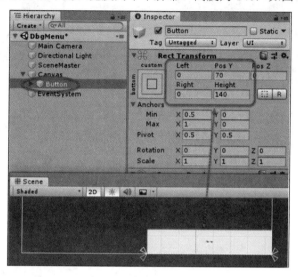

图 8-5

选中按钮下的 Text 游戏对象，修改 Text 文本属性为"退出"，修改 Font 字体属性为导入的字体文件；选中 Best Fit（最佳适应选项），修改 Max Size 最大尺寸为 100，如图 8-6 所示。

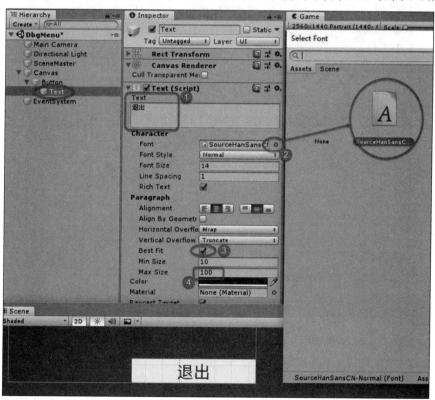

图 8-6

修改按钮名称为 ButtonExit，如图 8-7 所示。

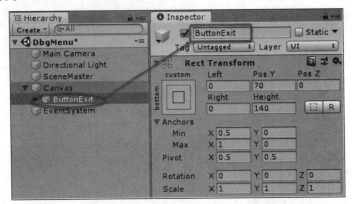

图 8-7

在 Game 窗口中可以选择预览的分辨率，如图 8-8 所示。

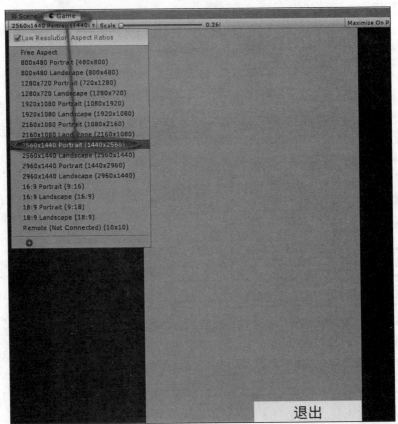

图 8-8

（3）添加修改其他按钮

复制并修改原有的按钮。添加地图按钮和删除地图按钮放在同一个位置，因为同一个时间只显示其中一个。按钮文本位置如表 8-1 所示。

表 8-1 按钮文本位置

Name（名称）	Text（文本）	Rect Transform（矩阵变化）（Left, Right, Pos Y, Height, Pos Z）	Anchors（锚点）（Min x, y, Max x, y）
ButtonExit	退出	0, 0, 70, 140, 0	0.5, 0, 1, 0
ButtonNavi	导航	0, 0, 70, 140, 0	0, 0, 0.5, 0
ButtonPoint	关键点	0, 0, 210, 140, 0	0, 0, 0.5, 0
ButtonRoad	路径	0, 0, 210, 140, 0	0.5, 0, 1, 0
ButtonModel	模型	0, 0, 350, 140, 0	0.5, 0, 1, 0
ButtonAddMap	添加地图	0, 0, 350, 140, 0	0, 0, 0.5, 0
ButtonDelMap	删除地图	0, 0, 350, 140, 0	0, 0, 0.5, 0

效果如图 8-9 所示。

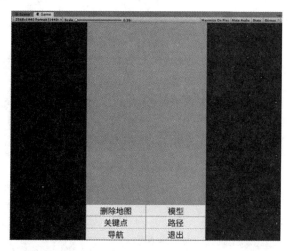

图 8-9

（4）添加设置文本输入框

单击菜单 GameObject→UI→Input Field，添加一个文本输入框，如图 8-10 所示。

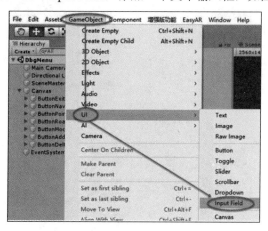

图 8-10

设置其对齐方式是底部对齐，Rect Transform 属性中的"Left，Right，Pos Y，Height，Pos Z"为"0，0，490，140，0"，使其在按钮上方，如图 8-11 所示。

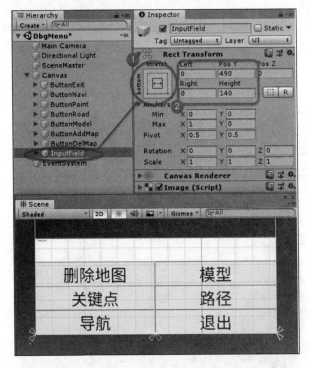

图 8-11

选中其下的 PlaceHolder 和 Text 游戏对象，修改 Font 字体属性为导入的字体文件，选中 Best Fit（最佳适应选项），修改 Max Size（最大尺寸）为 100，如图 8-12 所示。

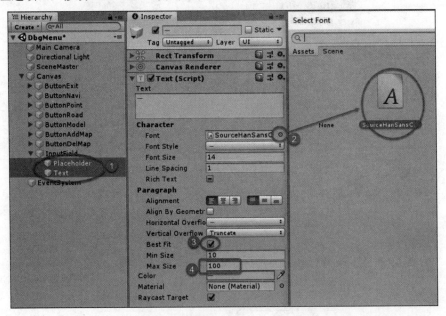

图 8-12

8.1.2 添加游戏控制脚本

在场景中添加一个空的游戏对象,并命名为 GameMaster。在 Kanamori/Scripts 目录下新建脚本并命名为 GameController,将其拖到 GameMaster 游戏对象上成为其组件,如图 8-13 所示。

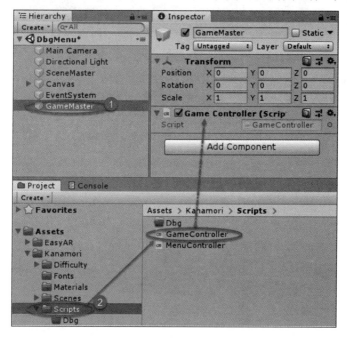

图 8-13

编辑脚本,添加命名空间。凡是实际场景中用到的脚本都在 Kanamori 命名空间下,预备场景专业脚本都在 Kanamori.Dbg 命名空间下。

```
namespace Kanamori
```

为了让 GameController 脚本能在每个场景中出现,使用 DontDestoryOnLoad 方法使其不会被卸载。为了避免场景中出现多个 GameController 脚本,使用单实例的方法来保证。

```
private static GameController instance = null;
void Awake()
{
    if (instance == null)
    {
        instance = this;
        DontDestroyOnLoad(gameObject);
    }
    else if (this != instance)
    {
        Destroy(gameObject);
        return;
    }
}
```

稀疏空间地图的读取和保存在 2 个场景中出现过，所以把功能写在 GameController 脚本中，统一读取、保存和删除。

```csharp
public string GetMapName()
{
    return PlayerPrefs.GetString("MapName", "");
}
public void SaveMapName(string mapName)
{
    PlayerPrefs.SetString("MapName", mapName);
}
public string GetMapID()
{
    return PlayerPrefs.GetString("MapID", "");
}
public void SaveMapID(string mapID)
{
    PlayerPrefs.SetString("MapID", mapID);
}
public void DelMap(){
    PlayerPrefs.DeleteKey("MapID");
    PlayerPrefs.DeleteKey("MapName");
}
```

8.1.3　修改设置场景控制脚本

（1）添加设置简单事件

在 MenuController 脚本中，添加退出和场景切换的方法。

```csharp
public void Quit()
{
    Application.Quit();
}
public void LoadScene(string sceneName)
{
    SceneManager.LoadScene(sceneName);
}
```

为了能测试场景切换，在 Kanamori/Scenes/Dbg 目录下，新建其他方便调试用的场景。单击菜单 File→Build Settings，将目录下的场景都加入到 Scenes In Build 中，并且确保 DbgMenu 是第一个场景，如图 8-14 所示。

选中 ButtonExit 游戏对象，单击"On Click()"下的"+"按钮，添加一个单击响应；将 SceneMaster 游戏对象拖到其下，设置单击响应的方法是 MenuController 脚本下的 Quit 方法，如图 8-15 所示。

用同样的方法设置 ButtonNavi 的 On Click()单击响应，把它设置为 SceneMaster 游戏对象下 MenuController 脚本下的 LoadScene 方法。因为该方法有输入值，设置输入值为 DbgNavigation，即要跳转到的场景名称，如图 8-16 所示。

第 8 章 调试场景开发

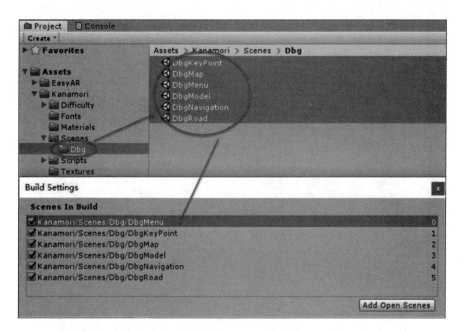

图 8-14

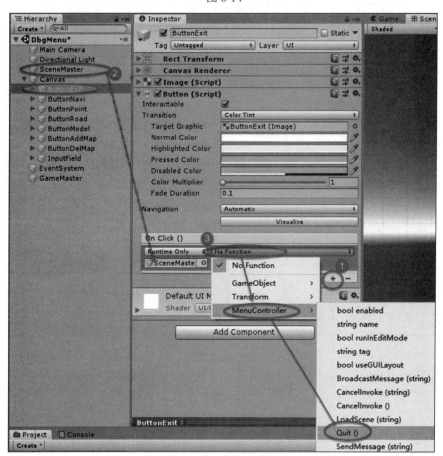

图 8-15

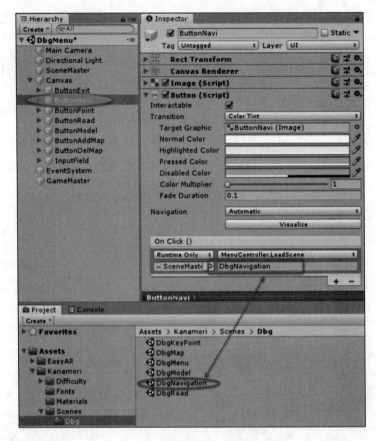

图 8-16

用同样的方法设置其他按钮，按钮方法及参数如表 8-2 所示。

表 8-2　按钮方法及参数

按钮名称	方　　法	参　　数
ButtonExit	Quit	
ButtonNavi	LoadScene	DbgNavigation
ButtonPoint	LoadScene	DbgKeyPoint
ButtonRoad	LoadScene	DbgRoad
ButtonModel	LoadScene	DbgModel
ButtonAddMap	LoadScene	DbgMap

（2）添加显示控制方法

添加地图、删除地图和输入框需要根据是否有地图进行控制，在场景载入和按钮单击后都需要控制。所以，新建方法控制界面，然后在其他地方调用。

```
private void UISettings()
{
    if (!game)
    {
```

```
            return;
        }
        if (game.GetMapID().Length == 0)
        {
            btnAdd.gameObject.SetActive(true);
            btnDel.gameObject.SetActive(false);
            input.text = "";
            input.interactable = true;
        }
        else
        {
            btnAdd.gameObject.SetActive(false);
            btnDel.gameObject.SetActive(true);
            input.text = game.GetMapName();
            input.interactable = false;
        }
    }
```

选中 SceneMaster 游戏对象，将 ButtonAddMap 游戏对象拖到 Btn Add 属性中为其赋值。将 ButtonDelMap 游戏对象拖到 Btn Del 属性中为其赋值。将 InputField 游戏对象拖到 Input 属性中为其赋值，如图 8-17 所示。

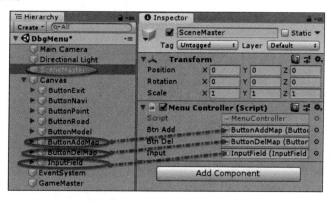

图 8-17

（3）添加设置删除地图按钮方法

```
public void DeleteMap()
{
    if (game)
    {
        game.DelMap();
        UISettings();
    }
}
```

设置 ButtonDelMap 的 On Click() 单击响应，把它设置为 SceneMaster 游戏对象下的 MenuController 脚本下的 DeleteMap 方法。

此时单击运行,所有按钮都能正确跳转。退出按钮单击无反应是因为退出在 Unity3D 编辑器中无效,但不影响实际使用。

(4)修改添加地图方法

添加地图的时候,需要加一个验证,避免地图名称为空。

```
public void AddMap(string sceneName)
{
    if (input.text.Length > 0)
    {
        LoadScene(sceneName);
    }
}
```

重新设置 ButtonAddMap 的 On Click()单击响应,把它设置为 SceneMaster 游戏对象下 MenuController 脚本下的 AddMap 方法,参数是 DbgMap。

菜单场景完成,MenuController 完整的代码如下:

```
using UnityEngine;
using UnityEngine.SceneManagement;
using UnityEngine.UI;

namespace Kanamori
{
    public class MenuController : MonoBehaviour
    {
        public Button btnAdd;
        public Button btnDel;
        public InputField input;
        private GameController game;
        void Start()
        {
            game = FindObjectOfType<GameController>();
            UISettings();
        }
        public void Quit()
        {
            Application.Quit();
        }
        public void LoadScene(string sceneName)
        {
            SceneManager.LoadScene(sceneName);
        }
        public void DeleteMap()
        {
            if (game)
            {
```

```
            game.DelMap();
            UISettings();
        }
    }
    public void AddMap(string sceneName)
    {
        if (input.text.Length > 0)
        {
            LoadScene(sceneName);
        }
    }
    private void UISettings()
    {
        if (!game)
        {
            return;
        }
        if (game.GetMapID().Length == 0)
        {
            btnAdd.gameObject.SetActive(true);
            btnDel.gameObject.SetActive(false);
            input.text = "";
            input.interactable = true;
        }
        else
        {
            btnAdd.gameObject.SetActive(false);
            btnDel.gameObject.SetActive(true);
            input.text = game.GetMapName();
            input.interactable = false;
        }
    }
}
```

8.2 地图场景开发

8.2.1 场景设置

（1）复制按钮

打开DbgMenu场景，选中Canvas和EventSystem游戏对象，单击鼠标右键，在弹出的菜单中选中Copy，复制UI画布和事件，如图8-18所示。

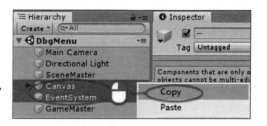

图 8-18

打开 DbgMap 场景，粘贴画布和事件。删除不用的按钮和文本输入框，留下最底部的按钮。修改按钮名称和文本，ButtonBack 按钮的文本是"返回"，ButtonSave 按钮的文本是"保存地图"，如图 8-19 所示。

图 8-19

（2）添加面板

添加面板的目的是让文本中的文字显示得更明显，不会因为背景而看不清。

单击菜单 GameObject→UI→Panel，往场景中添加一个面板，如图 8-20 所示。

选中 Panel 游戏对象，设置其 Anchor presets 锚点预设为 bottom-stretch（底部伸展），设置其 Rect Transform 矩阵变换"Left，Right，Pos Y，Height，Pos Z"为"0，0，210，140，0"，如图 8-21 所示。

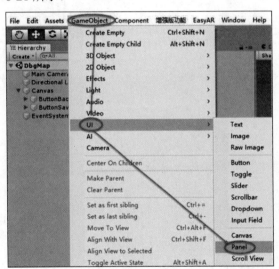

图 8-20

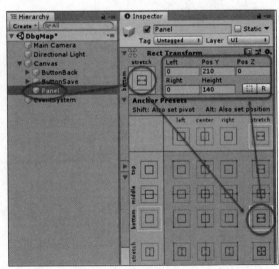

图 8-21

设置其 Image 组件的 Color 属性，使其透明度降低，如图 8-22 所示。

（3）添加文本

选中 Panel 游戏对象，单击鼠标右键，在弹出的菜单中选择 UI→Text，在 Panel 游戏对象中添加一个文本，如图 8-23 所示。

选中 Text 游戏对象，设置其 Anchor Presets 锚点预设为 stretch-stretch，与父游戏对象对齐，设置其 Rect Transform 矩阵变换"Left，Right，Top，Bottom，Pos Z"为"0，0，0，0，0"，即和父游戏对象一样大，如图 8-24 所示。

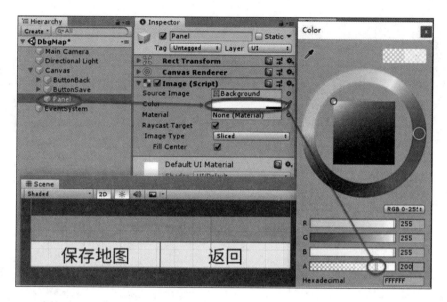

图 8-22

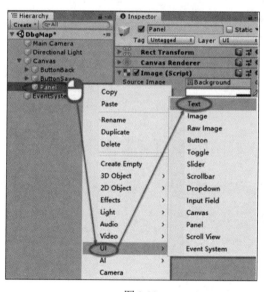

图 8-23

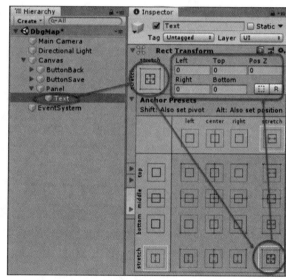

图 8-24

选中 Text 游戏对象，清空 Text 文本属性。修改 Font 字体属性为导入的字体文件，设置字体为水平居中对齐。选中 Best Fit（最佳适应选项），修改 Max Size（最大尺寸）为 100。

（4）添加脚本

因为新建稀疏空间地图场景在 Uinty3D 编辑器调试使用的和在移动端应用实际使用的不是同一个脚本，所以需要在 Kanamori/Scripts/Dbg 目录下建立脚本，如图 8-25 所示。

新建一个空的游戏对象并命名为 SceneMaster。在 Kanamori/Scripts/Dbg 目录下新建脚本 DbgMapController。将 DbgMapController 脚本拖到 SceneMaster 游戏对象上成为其组件，如图 8-26 所示。

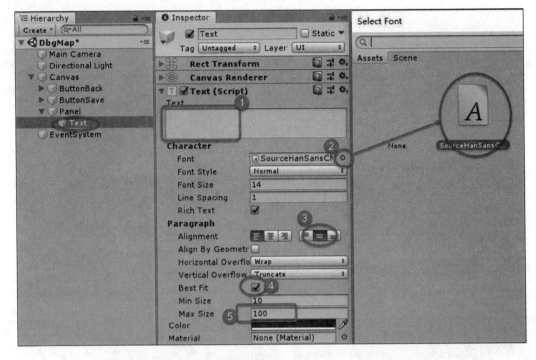

图 8-25

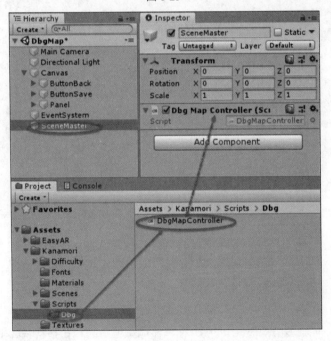

图 8-26

8.2.2 编写代码

（1）编写返回内容

因为返回菜单在多个场景中都存在，所以把具体功能放到 GameController 脚本中。

添加返回调试用菜单场景方法。

```
public void BackDbgMenu(){
    SceneManager.LoadScene("DbgMenu");
}
```

在 SceneController 脚本中添加返回方法，调用游戏控制里的方法实现。

```
public void Back()
{
    if (game)
    {
        game.BackDbgMenu();
    }
}
```

选中 ButtonBack 游戏对象，单击 On Click()下的"+"按钮，添加一个单击响应。将 SceneMaster 游戏对象拖到其下，设置单击响应的方法是 DbgMapController 脚本下的 Back 方法，如图 8-27 所示。

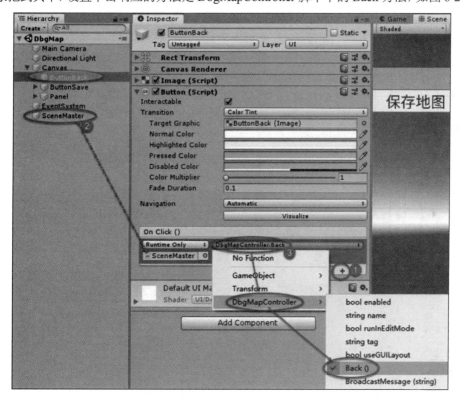

图 8-27

（2）编写保存内容

要保存的场景名称是从菜单场景带过来的，需要通过 GameController 来传递一下。
在 GameController 中增加一个变量：

```
public string inputName;
```

修改 MenuController 脚本中的方法，在添加地图的时候将文本输入框的内容赋值给 GameController 脚本的变量。

```
public void AddMap(string sceneName)
{
    if (input.text.Length > 0 && game)
    {
        game.inputName = input.text;
        LoadScene(sceneName);
    }
}
```

在 DbgMapController 脚本中添加用于显示的文本变量和保存地图的方法。

```
public Text text;
public void SaveMap()
{
    if (game)
    {
        game.SaveMapID("DbgMapTest");
        game.SaveMapName(game.inputName);
        text.text="地图保存成功。";
    }
}
```

选中 SceneMaster，将 Text 游戏对象拖到 Text 属性中为其赋值，如图 8-28 所示。

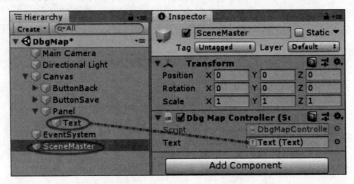

图 8-28

DbgMapController 脚本完整内容如下：

```
using UnityEngine;
using UnityEngine.UI;

namespace Kanamori.Dbg
{
    public class DbgMapController : MonoBehaviour
    {
        private GameController game;
```

```
        public Text text;
        void Start()
        {
            game = FindObjectOfType<GameController>();
        }
        public void SaveMap()
        {
            if (game)
            {
                game.SaveMapID("DbgMapTest");
                game.SaveMapName(game.inputName);
                text.text="地图保存成功。";
            }
        }
        public void Back()
        {
            if (game)
            {
                game.BackDbgMenu();
            }
        }
    }
}
```

此时运行，所有场景调试都必须从菜单场景开始运行才能正确。输入地图名称，单击"添加地图"按钮，正确跳转，如图 8-29 所示。

单击"保存地图"按钮，正确保存并提示；单击"返回"按钮返回菜单，如图 8-30 所示。如果已经保存过地图，就会显示"删除地图"按钮和地图名称，如图 8-31 所示。

图 8-29　　　　　　　　　图 8-30　　　　　　　　　图 8-31

8.3 模型场景开发

往场景中动态添加模型，可以像官方例子那样，通过拖动一个 UI 来实现。这样的实现方式对初学者而言稍微有点麻烦，所以这里使用了一个简单的实现方式，单击按钮以后在屏幕前方放置一个模型。

8.3.1 模型移动功能预制件开发

在模型场景中，选中模型后，进行控制的功能相对麻烦，但是可以单独独立出来。所以，将这部分内容独立出来，做成一个 Prefab 预制件。

1. 设置界面

（1）添加面板

打开 DbgModel 场景（也可以打开一个空的场景，不影响最终结果）。单击菜单 GameObject→UI→Panel，往场景中添加一个面板。

选中 Panel 游戏对象，设置其 Anchor Presets 锚点预设为 bottom-stretch（底部伸展），设置其 Rect Transform 矩阵变换 "Left，Right，Pos Y，Height，Pos Z" 为 "0，0，225，450，0"，修改其名称为 PanelController，如图 8-32 所示。

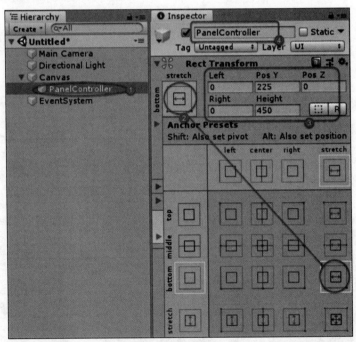

图 8-32

（2）添加左边按钮

选中 PanelController 游戏对象，单击鼠标右键，在弹出的菜单中选择 UI→Button，在 PanelController 游戏对象下添加一个按钮，如图 8-33 所示。

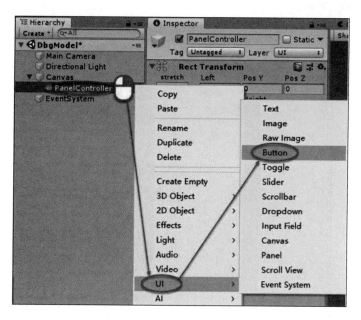

图 8-33

选中 Button 游戏对象，设置其 Anchor Presets 锚点预设为 bottom-left（左下角），设置其 Rect Transform 矩阵变换"Pos X，Width，Pos Y，Height，Pos Z"为"75，150，75，150，0"，修改其名称为 ZSub，如图 8-34 所示。

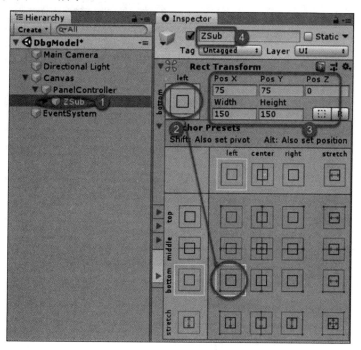

图 8-34

选中 ZSub 游戏对象下的 Text 游戏对象，设置 Text 文本属性为"Z-"；选中 Best Fit（最佳适应选项），修改 Max Size（最大尺寸）为 100，如图 8-35 所示。

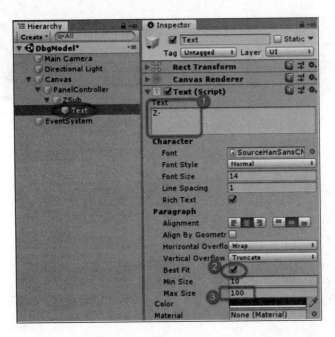

图 8-35

复制并修改原有的按钮，总共需要 6 个按钮，如图 8-36 所示。按钮文本和位置如表 8-3 所示。

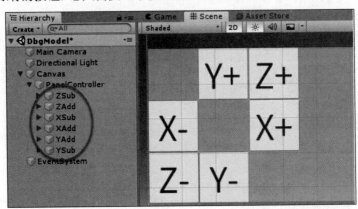

图 8-36

表 8-3　按钮文本和位置

Name（名称）	Text（文本）	Rect Transform（矩阵变化） （Pos X，Width，Pos Y，Height，Pos Z）	说　明
ZSub	Z-	75，150，75，150，0	Z 轴减少
ZAdd	Z+	375，150，375，150，0	Z 轴增加
XSub	X-	75，150，225，150，0	X 轴减少
XAdd	X+	375，150，225，150，0	X 轴增加
YAdd	Y+	225，150，375，150，0	Y 轴增加
YSub	Y-	225，150，75，150，0	Y 轴减少

（3）添加右边按钮

选中 PanelController 游戏对象，单击鼠标右键，在弹出的菜单中选择 UI→Button，在 PanelController 游戏对象下添加一个按钮 Button。

选中 Button 游戏对象，把其 Anchor Presets 锚点预设为 bottom-right（右下角），设置其"Rect Transform"矩阵变换"Pos X，Width，Pos Y，Height，Pos Z"为"-75，150，75，150，0"，修改其名称为"Scale"，如图 8-37 所示。

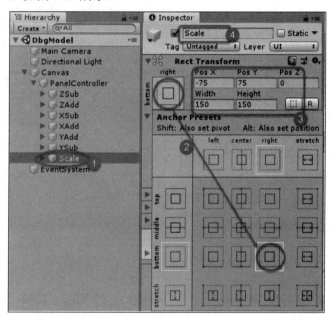

图 8-37

选中 Scale 游戏对象下的 Text 游戏对象，设置 Text 文本属性为"Z-"，选中 Best Fit（最佳适应选项），修改 Max Size（最大尺寸）为 100。

复制并修改原有的按钮，总共需要 6 个按钮。按钮文本和位置如表 8-4 所示。

表 8-4　按钮文本和位置

Name（名称）	Text（文本）	Rect Transform（矩阵变化） （Pos X，Width，Pos Y，Height，Pos Z）	说　明
Scale	Scale	-75，150，75，150，0	控制缩放
Rotation	Rotation	-75，150，225，150，0	控制角度
Position	Position	-75，150，375，150，0	控制位置
Ten	10	-225，150，375，150，0	增减 10
One	1	-225，150，225，150，0	增减 1
OneTenth	0.1	22-225，150，75，150，0	增减 0.1

（4）添加显示信息文本

选中 PanelController 游戏对象，单击鼠标右键，在弹出的菜单中选择 UI→Text，在 PanelController 游戏对象下添加一个文本，如图 8-38 所示。

选中 Text 游戏对象，设置其 Anchor Presets 锚点预设为 bottom-stretch（底部拉伸），设置其 Rect Transform 矩阵变换"Left，Right，Pos Y，Height，Pos Z"为"450，300，225，450，0"，如图 8-39 所示。这样设置以后，Text 游戏对象的宽度是跟随父节点改变的，其他按钮大小不变。屏幕宽度在一定程度上只会影响 Text 游戏对象的宽度。

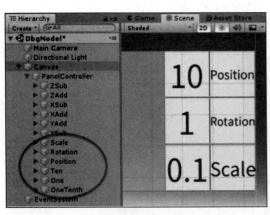
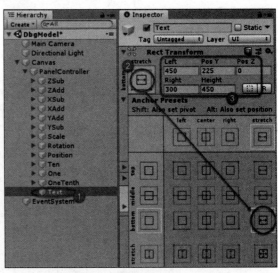

图 8-38　　　　　　　　　　　　　　　　图 8-39

选中 Text 游戏对象，修改 Text 文本属性为显示时候的情况；设置 Alignment（对齐方式）为垂直居中对齐；选中 Align By Geometry（几何对齐选项）、Best Fit（最佳适应选项），修改 Max Size 为 100，如图 8-40 所示。

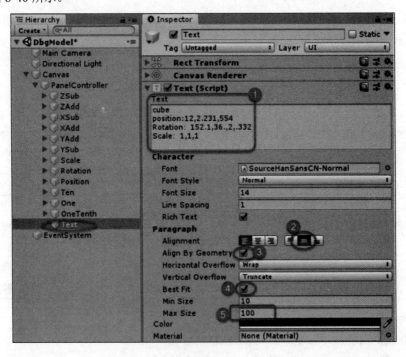

图 8-40

2. 添加编写脚本

（1）添加脚本

在 Kanamori/Scripts 目录下添加脚本 UIControlObject，并将其拖到 PanelController 游戏对象下成为其组件，如图 8-41 所示。

（2）按钮分组添加事件

按钮很多，按照使用方法分成 3 个组来添加。在代码中添加 3 个公开的按钮数组。

```
public Button[] btnActions;
public Button[] btnTypes;
public Button[] btnNumbers;
public Text txtShow;
```

在 Unity3D 编辑器中，设置 3 个组的大小和具体内容。其中，Btn Numbers 数组大小为 3，包括 Ten、One 和 OneTenth；Btn Types 数组大小为 3，包括 Scale、Rotation 和 Position；其余 6 个按钮都在 Btn Actions 中，如图 8-42 所示。数组中的顺序不重要。

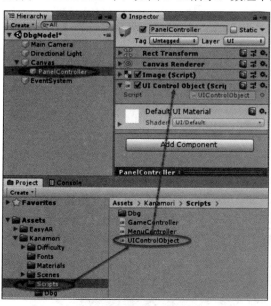

图 8-41

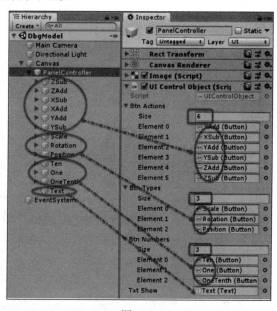

图 8-42

添加 3 种类型的响应事件，通过按钮的名称区分具体按的是哪一个。在 Start 方法中用遍历来给每个按钮添加单击事件响应。

```
void Start()
{
    foreach (Button btn in btnNumbers)
    {
        btn.onClick.AddListener(delegate ()
        {
            OnClickNumber(btn.name);
        });
```

```
    }
    foreach (var btn in btnTypes)
    {
        btn.onClick.AddListener(delegate ()
        {
            OnClickType(btn.name);
        });
    }
    foreach (var btn in btnActions)
    {
        btn.onClick.AddListener(delegate ()
        {
            OnClickAction(btn.name);
        });
    }
}
private void OnClickNumber (string btnName)
{
    Debug.Log(btnName);
}
private void OnClickType(string btnName)
{
    Debug.Log(btnName);
}
private void OnClickAction(string btnName)
{
    Debug.Log(btnName);
}
```

可以在控制台上看一下是否正确，如图 8-43 所示。

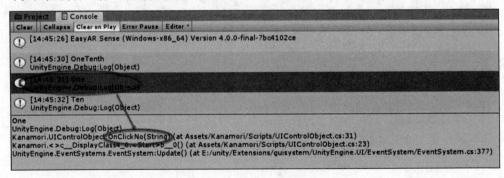

图 8-43

（3）设置按钮颜色

为了让使用者知道当前是什么操作，对类型和数值按钮组设置颜色显示。

添加 2 个公开颜色属性，这样使用者能自己设置具体的颜色。因为在多个地方用到，所以将其抽取成一个单独的函数来实现。

```
public Color colorActive;
public Color colorInactive;
void Start()
{
    ...
    SetButtonColor(btnNumbers, "One");
    SetButtonColor(btnTypes, "Position");
}
private void OnClickNumber (string btnName)
{
    Debug.Log(btnName);
    SetButtonColor(btnNos, btnName);
}
private void OnClickType(string btnName)
{
    Debug.Log(btnName);
    SetButtonColor(btnTypes, btnName);
}
private void SetButtonColor(Button[] buttons, string btnName)
{
    foreach (var btn in buttons)
    {
        Color temp = colorInactive;
        if (btn.name.Equals(btnName))
        {
            temp = colorActive;
        }
        btn.GetComponent<Image>().color = temp;
    }
}
```

这样，在脚本上设置颜色以后，单击具体的按钮，单击过的按钮和其他颜色不一样，就有选中的感觉，如图 8-44 所示。

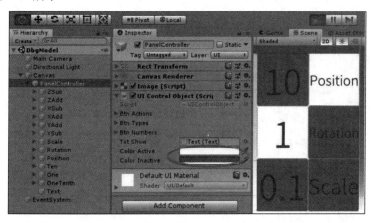

图 8-44

（4）对游戏对象进行控制

数值组按钮和类型组按钮点击的结果放到 2 个属性中，为了调试方便，均设置为公开属性。

```
public string actionType;
public float actionNumber;
void Start()
{
   ...
   actionType = "Position";
   actionNumber = 1;
}
private void OnClickNumber(string btnName)
{
   Debug.Log(btnName);
   SetButtonColor(btnNumbers, btnName);
   switch (btnName)
   {
      case "Ten":
         actionNumber = 10;
         break;
      case "One":
         actionNumber = 1;
         break;
      case "OneTenth":
         actionNumber = 0.1f;
         break;
   }
}
private void OnClickType(string btnName)
{
   Debug.Log(btnName);
   SetButtonColor(btnTypes, btnName);
   actionType=btnName;
}
```

单击动作按钮组的按钮，根据数值组按钮和类型组按钮单击的结果，对选中的游戏对象进行操作。

```
public Transform selected;
private void OnClickAction(string btnName)
{
   Debug.Log(btnName);
   if (selected)
   {
      var temp = Vector3.zero;
      switch (btnName)
      {
```

```
            case "XAdd":
                temp = new Vector3(actionNumber, 0, 0);
                break;
            case "XSub":
                temp = new Vector3(-actionNumber, 0, 0);
                break;
            case "YAdd":
                temp = new Vector3(0, actionNumber, 0);
                break;
            case "YSub":
                temp = new Vector3(0, -actionNumber, 0);
                break;
            case "ZAdd":
                temp = new Vector3(0, 0, actionNumber);
                break;
            case "ZSub":
                temp = new Vector3(0, 0, -actionNumber);
                break;
        }
        switch (actionType)
        {
            case "Position":
                selected.localPosition = selected.localPosition + temp;
                break;
            case "Rotation":
                selected.localEulerAngles = selected.localEulerAngles + temp;
                break;
            case "Scale":
                selected.localScale = selected.localScale + temp;
                break;
        }
    }
}
```

（5）设置显示文本

用 string.Format 方法拼接字符串。

```
public void SetSelected(Transform tfSelected)
{
    selected = tfSelected;
    ShowSelectedInfo();
}
public void ClearSelected()
{
    selected = null;
    txtShow.text = "none";
}
```

```
private void ShowSelectedInfo()
{
    if (!selected)
    {
        return;
    }
    txtShow.text=string.Format(
        @"Name:{0}
        Position:{1}
        Rotation:{2}
        Scale:{3}",
        selected.name,
        selected.localPosition,
        selected.localEulerAngles,
        selected.localScale).Replace(" ","");
}
```

（6）生成预制件

在目录Kanamori下新建一个目录并命名为Prefabs，将PanelController游戏对象拖到该目录中成为预制件，如图8-45所示。删除场景中添加的内容。

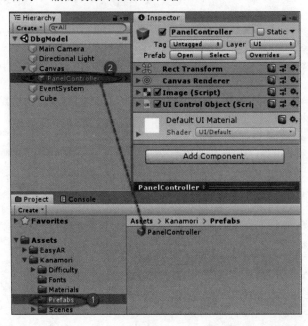

图 8-45

UIControlObject 完整的代码如下：

```
using UnityEngine;
using UnityEngine.UI;
namespace Kanamori
{
    public class UIControlObject : MonoBehaviour
```

```csharp
{
    public Button[] btnActions;
    public Button[] btnTypes;
    public Button[] btnNumbers;
    public Text txtShow;
    public Color colorActive;
    public Color colorInactive;
    public string actionType;
    public float actionNumber;
    public Transform selected;
    void Start()
    {
        foreach (Button btn in btnNumbers)
        {
            btn.onClick.AddListener(delegate ()
            {
                OnClickNumber(btn.name);
            });
        }
        foreach (var btn in btnTypes)
        {
            btn.onClick.AddListener(delegate ()
            {
                OnClickType(btn.name);
            });
        }
        foreach (var btn in btnActions)
        {
            btn.onClick.AddListener(delegate ()
            {
                OnClickAction(btn.name);
            });
        }
        SetButtonColor(btnNumbers, "One");
        SetButtonColor(btnTypes, "Position");
        actionType = "Position";
        actionNumber = 1;
        ClearSelected();
    }
    private void OnClickNumber(string btnName)
    {
        SetButtonColor(btnNumbers, btnName);
        switch (btnName)
        {
            case "Ten":
                actionNumber = 10;
                break;
```

```csharp
                case "One":
                    actionNumber = 1;
                    break;
                case "OneTenth":
                    actionNumber = 0.1f;
                    break;
            }
        }
        private void OnClickType(string btnName)
        {
            SetButtonColor(btnTypes, btnName);
            actionType = btnName;
        }
        private void OnClickAction(string btnName)
        {
            if (selected)
            {
                var temp = Vector3.zero;
                switch (btnName)
                {
                    case "XAdd":
                        temp = new Vector3(actionNumber, 0, 0);
                        break;
                    case "XSub":
                        temp = new Vector3(-actionNumber, 0, 0);
                        break;
                    case "YAdd":
                        temp = new Vector3(0, actionNumber, 0);
                        break;
                    case "YSub":
                        temp = new Vector3(0, -actionNumber, 0);
                        break;
                    case "ZAdd":
                        temp = new Vector3(0, 0, actionNumber);
                        break;
                    case "ZSub":
                        temp = new Vector3(0, 0, -actionNumber);
                        break;
                }
                switch (actionType)
                {
                    case "Position":
                        selected.localPosition = selected.localPosition + temp;
                        break;
                    case "Rotation":
                        selected.localEulerAngles = selected.localEulerAngles + temp;
                        break;
```

```csharp
                    case "Scale":
                        selected.localScale = selected.localScale + temp;
                        break;
                }
            }
            ShowSelectedInfo();
        }
        private void SetButtonColor(Button[] buttons, string btnName)
        {
            foreach (var btn in buttons)
            {
                Color temp = colorInactive;
                if (btn.name.Equals(btnName))
                {
                    temp = colorActive;
                }
                btn.GetComponent<Image>().color = temp;
            }
        }
        public void SetSelected(Transform tfSelected)
        {
            selected = tfSelected;
            ShowSelectedInfo();
        }
        public void ClearSelected()
        {
            selected = null;
            txtShow.text = "none";
        }
        private void ShowSelectedInfo()
        {
            if (!selected)
            {
                return;
            }
            txtShow.text=string.Format(
                @"Name:{0}
                Position:{1}
                Rotation:{2}
                Scale:{3}",
                selected.name,
                selected.localPosition,
                selected.localEulerAngles,
                selected.localScale).Replace(" ","");
        }
    }
}
```

8.3.2 场景设置

1. 添加界面切换对象

打开 DbgModel 场景，删除之前添加的内容，使其成为一个默认的场景。单击菜单 GameObject→UI→Canvas，添加一个画布，如图 8-46 所示。

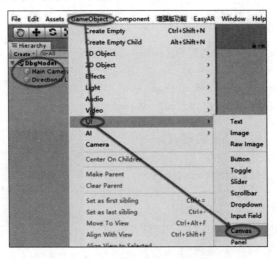

图 8-46

选中 Canvas 游戏对象，单击鼠标右键，在弹出的菜单中选中 Create Empty，添加空的游戏对象，总共添加 2 个空的游戏对象。

选中新增的游戏对象，设置其 Anchor Presets 锚点预设为 stretch–stretch（屏幕对齐），设置其 Rect Transform 矩阵变换"Left，Right，Top，Bottom，Pos Z"为"0，0，0，0，0"，修改其名称为 AddUI 和 SaveUI，如图 8-47 所示。这两个游戏对象用于放置 2 部分界面，通过对这两个游戏对象的禁用和启用实现界面切换。

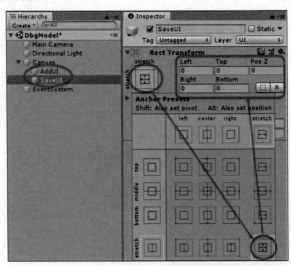

图 8-47

2. 添加 UI

在 AddUI 游戏对象下，按照前面章节的方法添加两个按钮，如图 8-48 所示。按钮文本和位置如表 8-5 所示。

图 8-48

表 8-5 按钮文本和位置

Name（名称）	Text（文本）	Anchors（锚点） （Min X, Min Y, Max X, Max Y）	Rect Transform（矩阵变化）(Left, Right, Pos Y, Height, Pos Z)
ButtonBack	返回	0.5, 0, 1, 0	0, 0, 70, 140, 0
ButtonAdd	添加	0, 0, 0.5, 0	0, 0, 70, 140, 0

在 SaveUI 游戏对象下，按照前面章节的方法添加 UI，分别添加按钮、输入文本框以及在面板下的文本，如图 8-49 所示。按钮名称文本和位置如表 8-6 所示。

图 8-49

表 8-6 按钮名称文本和位置

类　型	Name（名称）	Text（文本）	Anchors（锚点）(Min X, Min Y, Max X, Max Y)	Rect Transform（矩阵变化）(Left, Right, Pos Y, Height, Pos Z)
按钮	ButtonClose	关闭	0.5, 0, 1, 0	0, 0, 70, 140, 0
按钮	ButtonSave	保存	0, 0, 0.5, 0	0, 0, 70, 140, 0
按钮	ButtonDelete	删除	0.5, 0, 1, 0	0, 0, 210, 140, 0
面板	Panel		0, 0, 0.5, 0	0, 0, 210, 140, 0
文本	Text		0, 0, 1, 1	0, 0, 0, 0, 0

3. 添加预制件

将前一节做的 Kanamori/Prefabs 目录下的 PanelController 拖到 SaveUI 游戏对象下，成为其子对象，设置其 Rect Transform 矩阵变换"Left，Right，Pos Y，Height，Pos Z"为"0，0，500，450，0"，如图 8-50 所示。

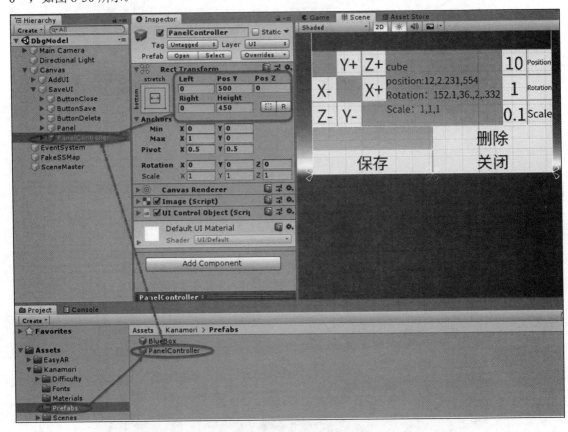

图 8-50

4. 添加虚拟的稀疏空间地图游戏对象

在场景中添加一个空的游戏对象并命名为 FakeSSMap，即在编辑器模拟的时候认为该游戏对象就是稀疏空间地图的游戏对象，如图 8-51 所示。

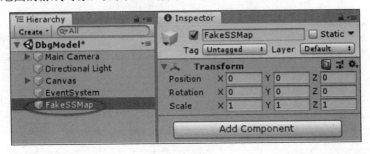

图 8-51

8.3.3 添加虚拟物体功能

1. 添加一个半透明的盒子

打开一个新的场景,单击菜单 GameObject→3D Object→Cube,为场景添加一个方块,设置方块的位置和角度都是默认值,拖一个白色的纹理到方块上,如图 8-52 所示。

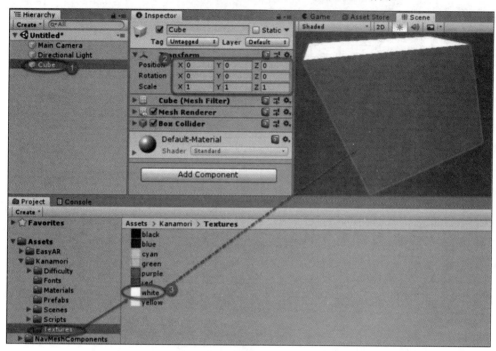

图 8-52

修改方块的名称为 BlueBox,修改 Rendering Mode 为 Transparent,如图 8-53 所示。

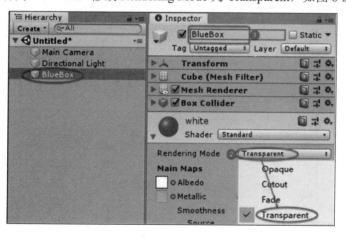

图 8-53

修改 Albedo 选项,将颜色修改为蓝色、透明值修改为 100,这样就变成了一个半透明的蓝色盒子,如图 8-54 所示。

将 BlueBox 游戏对象拖到 Kanamori/Prefabs 目录中成为 prefab 预制件，如图 8-55 所示。

图 8-54

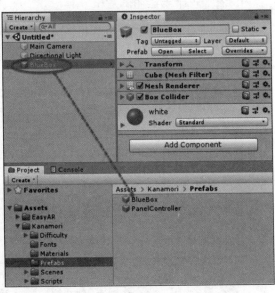

图 8-55

2. 添加生成点

屏幕前方的位置可以通过计算获得。为了简便起见，直接在摄像头前方放置一个物体，这样就不需要计算了。

选中 Main Camera 游戏对象，单击鼠标右键，在弹出的菜单中选择 3D Object→Sphere，添加一个球体作为 Main Camera 游戏对象的子对象。这里其实添加一个空的游戏对象就可以了，添加球体只是为了在需要调试的时候看上去方便。

设置 Sphere 游戏对象的 Position 值为 "0，0，1.5"，即在摄像机前方 1.5 米处。去掉 Mesh Renderer 和 Sphere Collider 组件的激活选项，这样就隐藏起来，不会影响到其他模型了，如图 8-56 所示。

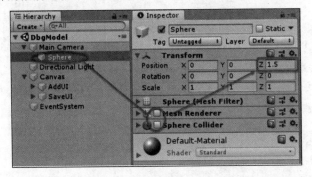

图 8-56

3. 添加并编写脚本

在场景中添加一个空的游戏对象并命名为 SceneMaster。在 Kanamori/Scripts/Dbg 目录下添加一个脚本 DbgModelController，并将其拖到 SceneMaster 游戏对象下成为其组件，如图 8-57 所示。

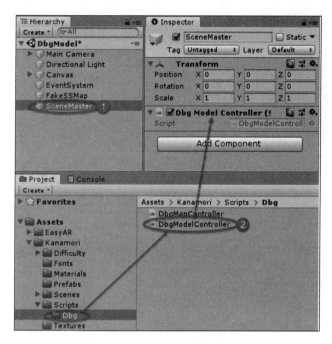

图 8-57

添加返回按钮的功能:

```
private GameController game;
void Start()
{
    game=FindObjectOfType<GameController>();
}
public void Back(){
    game.BackDbgMenu();
}
```

添加公开属性获取预制件和稀疏空间地图。用 Instantiate 方法在地图游戏对象下生成一个蓝色盒子,位置在摄像机前方,角度是默认值。

```
public Transform frontCamera;
public Transform ssMap;
public void Add(){
    Instantiate(blueBox,frontCamera.position,Quaternion.identity,ssMap);
}
```

选中 SceneMaster 游戏对象,将 Sphere 游戏对象拖到 Front Camera 属性中为其赋值,将 FakeSSMap 游戏对象拖到 Ss Map 属性中为其赋值,将 Kanamori/Prefabs 目录下的 BlueBox 预制件拖到 Blue Box 属性中为其赋值,如图 8-58 所示。

设置 ButtonBack 按钮的单击事件方法,把它设置为 DbgModelController 脚本下的 Back 方法,如图 8-59 所示。设置 ButtonAdd 按钮的单击事件方法为 DbgModelController 脚本下的 Add 方法。

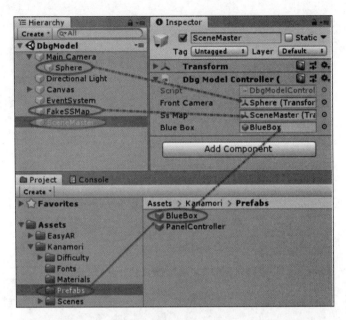

图 8-58

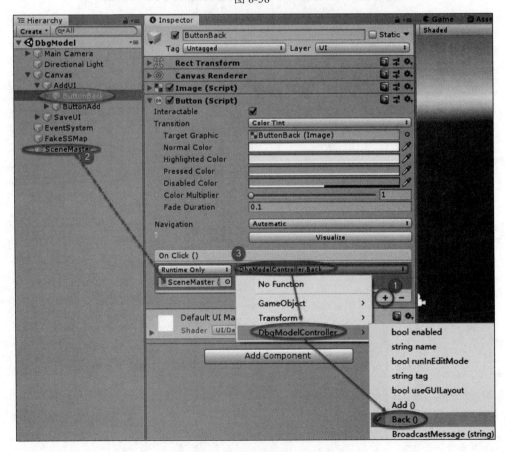

图 8-59

测试的时候，可以先暂时去掉 SaveUI 激活选项，如图 8-60 所示。

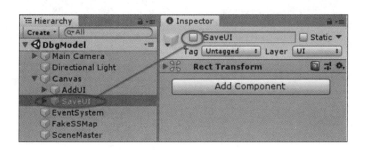

图 8-60

从 DbgMenu 场景开始运行,单击模型按钮进入当前场景;单击"返回"按钮就能返回菜单场景;单击"添加"按钮,就会在 FakeSSMap 游戏对象下添加方块,如图 8-61 所示。

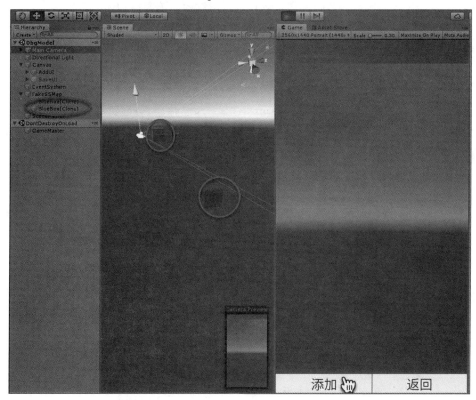

图 8-61

8.3.4 界面切换和点击选中

1. 界面切换

添加两套 UI 的游戏对象属性,用 SetActive 控制是否激活。

```
public GameObject addUI;
public GameObject saveUI;
void Start()
{
    game=FindObjectOfType<GameController>();
```

```
        addUI.SetActive(true);
        saveUI.SetActive(false);
    }
    public void Close(){
        addUI.SetActive(true);
        saveUI.SetActive(false);
    }
```

2. 点击切换

点击了放置的虚拟物体后，界面切换到保存的界面。

Unity3D 点击屏幕选中虚拟物体的过程是在点击屏幕的那个点发出一条射线。如果这条射线从 Unity3D 的摄像机发出后照射到虚拟物体，就认为点击选中了该虚拟物体。

在不同的环境下，点击方式不一样，有鼠标点击或者屏幕点击。此外，还要考虑 UI 击穿的问题，即当在屏幕上点击的是界面 UI 的时候不需要发出射线。

```
void Update()
{
    if (Application.platform == RuntimePlatform.WindowsEditor)
    {
        if (Input.GetMouseButtonDown(0)
        && !EventSystem.current.IsPointerOverGameObject())
        {
            Ray ray = Camera.main.ScreenPointToRay(Input.mousePosition);
            TouchedObject(ray);
        }
    }
    else
    {
        if (Input.touchCount == 1
        && Input.touches[0].phase == TouchPhase.Began
        && !EventSystem.current.IsPointerOverGameObject
(Input.touches[0].fingerId))
        {
            Ray ray = Camera.main.ScreenPointToRay(Input.touches[0].position);
            TouchedObject(ray);
        }
    }
}
private void TouchedObject(Ray ray)
{
    if (Physics.Raycast(ray, out RaycastHit hit))
    {
        addUI.SetActive(false);
        saveUI.SetActive(true);
    }
}
```

在不是很重要或者要求不严格的场合下,可以用鼠标点击代替屏幕点击。

```
void Update()
{
    if (Input.GetMouseButtonDown(0)
    && !EventSystem.current.IsPointerOverGameObject())
    {
        Ray ray = Camera.main.ScreenPointToRay(Input.mousePosition);
        if (Physics.Raycast(ray, out RaycastHit hit))
        {
            addUI.SetActive(false);
            saveUI.SetActive(true);
        }
    }
}
```

3. 设置关闭按钮

设置 ButtonClose 按钮的点击事件方法为 DbgModelController 脚本下的 Close 方法,如图 8-62 所示。

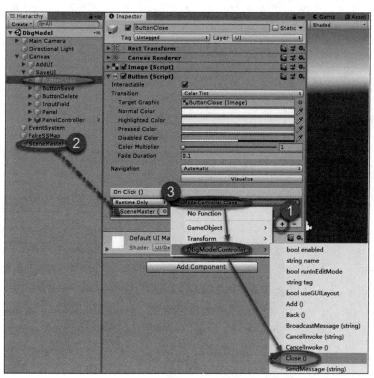

图 8-62

4. 设置控制

添加一个 UIControlObject 属性,在点击以后,用 SetSelected 方法设置其控制的游戏对象,就能利用先前写的 UI 来控制所选中游戏对象的位置、角度。

```
private UIControlObject uiControl;
void Start()
{
    ...
    uiControl=FindObjectOfType<UIControlObject>();
}
private void TouchedObject(Ray ray)
{
    if (Physics.Raycast(ray, out RaycastHit hit))
    {
        addUI.SetActive(false);
        saveUI.SetActive(true);
        uiControl.SetSelected(hit.transform);
    }
}
```

从 DbgMenu 场景开始运行，添加方块后，点击能够选中并且控制，如图 8-63 所示。

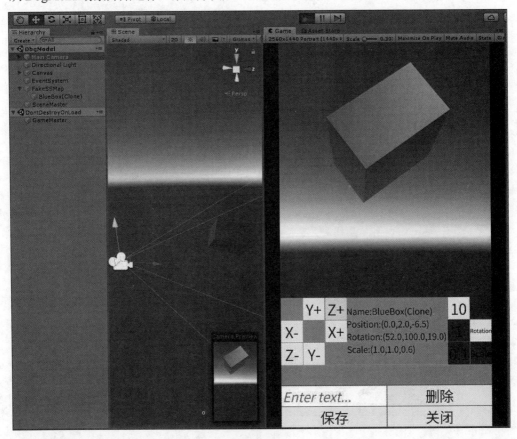

图 8-63

8.3.5 删除和保存

这里需要将动态添加的内容保存下来，我们可以使用之前的方法。

1. 添加简单类

在 Kanamori/Scripts 目录下添加一个脚本，命名为 DynamicObject，如图 8-64 所示。
该类很简单，不继承 MonoBehaviour，只有 3 个公开属性。

```
using UnityEngine;
namespace Kanamori
{
    public class DynamicObject
    {
        public Vector3 position;
        public Vector3 rotation;
        public Vector3 scale;
    }
}
```

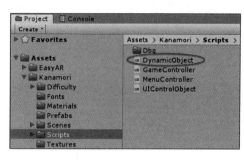

图 8-64

2. 修改 GameController 类添加保存方法

接下来修改 GameController 类，添加将字符串保存到文本以及从文本读取字符串的方法。这 2 个方法都是私有的，不能直接使用。

```
private void SaveStringArray(string[] stringArray, string path)
{
        try
        {
           using (StreamWriter writer = new StreamWriter(path))
           {
              foreach (var s in stringArray)
              {
                 writer.WriteLine(s);
              }
           }
        }
        catch (Exception ex)
        {
           Debug.Log(ex.Message);
        }
}
private List<string> LoadStringList(string path)
{
        List<string> list = new List<string>();
        try
        {
           using (StreamReader reader = new StreamReader(path))
           {
              while (!reader.EndOfStream)
              {
                 list.Add(reader.ReadLine());
```

```csharp
                }
            }
        }
        catch (Exception ex)
        {
            Debug.Log(ex.Message);
        }
        return list;
    }
```

添加一个只读属性保存路径，用公开方法进行读取和保存，以避免路径错误。

```csharp
private static readonly string pathDynamicObject = "/dynamicobject.txt";
public void SaveDynamicObject(string[] stringArray)
{
    SaveStringArray(stringArray, Application.persistentDataPath + pathDynamicObject);
}
public List<string> LoadDynamicObject()
{
    return LoadStringList(Application.persistentDataPath + pathDynamicObject);
}
```

3. 修改 DbgMapController 脚本添加保存和删除方法

保存的时候，遍历稀疏空间地图，将其子游戏对象的位置和角度信息保存下来。删除的时候，直接删除游戏对象即可。

```csharp
public void Save()
{
    string[] jsons = new string[ssMap.childCount];
    for (int i = 0; i < ssMap.childCount; i++)
    {
        DynamicObject dynamicObject = new DynamicObject();
        dynamicObject.position = ssMap.GetChild(i).localPosition;
        dynamicObject.rotation = ssMap.GetChild(i).localEulerAngles;
        dynamicObject.scale = ssMap.GetChild(i).localScale;
        jsons[i] = JsonUtility.ToJson(dynamicObject);
    }
    game.SaveDynamicObject(jsons);
    textShow.text = "保存" + ssMap.childCount + "个游戏对象";
}

public void Delete()
{
    var go = uiControl.selected.gameObject;
    uiControl.ClearSelected();
    Destroy(go);
```

```
        textShow.text = "删除选中物体，请保存结果。";
    }
```

4. 设置脚本

设置 ButtonSave 按钮的点击事件方法为 DbgModelController 脚本下的 Save 方法。设置 ButtonDelete 按钮的点击事件方法为 DbgModelController 脚本下的 Delete 方法，如图 8-65 所示。

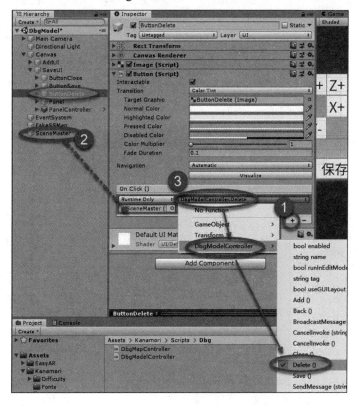

图 8-65

选中 SceneMaster 游戏对象，将面板下的 Text 游戏对象拖到 Text Show 属性中，并为其赋值，如图 8-66 所示。

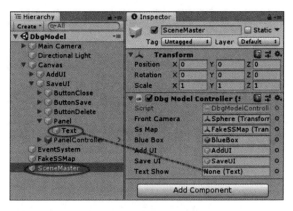

图 8-66

从菜单场景开始运行，就能保存添加的游戏对象了，如图 8-67 所示。

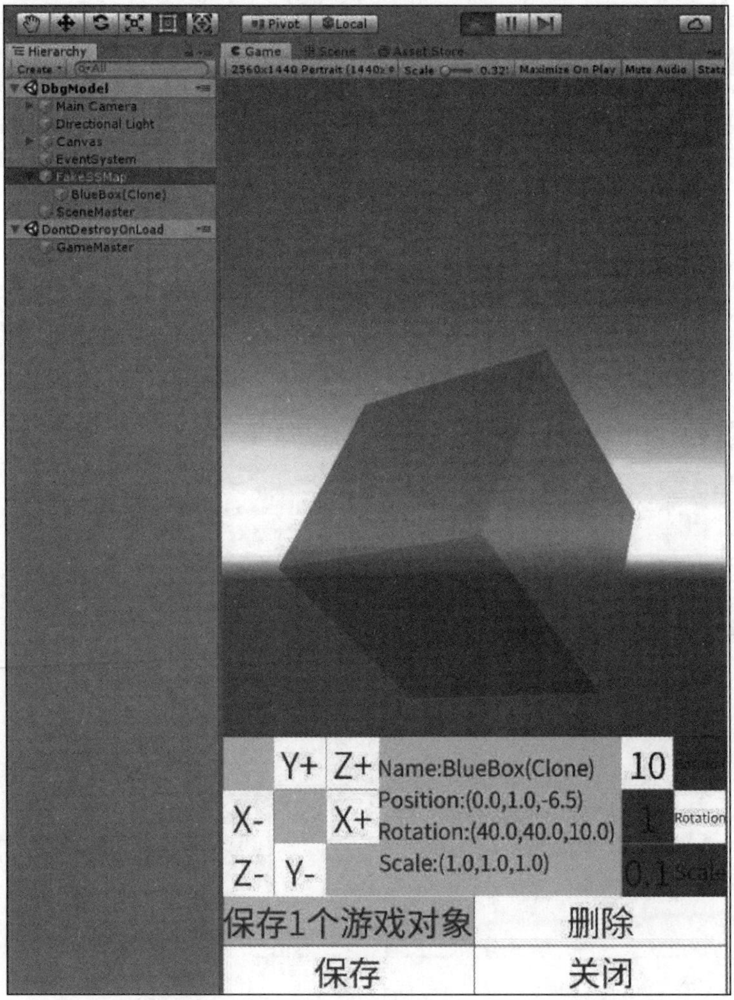

图 8-67

打开指定目录可以看到保存的文本文件及其内容，如图 8-68 所示。

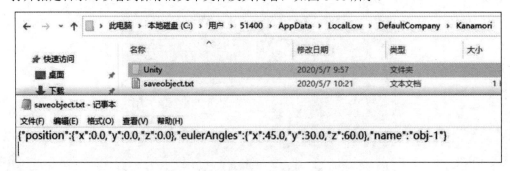

图 8-68

最后添加上加载功能，启动以后就能读取保存过的内容并生成对应的游戏对象。

```csharp
private void Load()
{
    var list = game.LoadDynamicObject();
    foreach (var item in list)
    {
        var dynamicObject = JsonUtility.FromJson<DynamicObject>(item);
        var tf = Instantiate(blueBox, ssMap).transform;
        tf.localPosition = dynamicObject.position;
        tf.localEulerAngles = dynamicObject.rotation;
        tf.localScale = dynamicObject.scale;
    }
}
```

脚本的完整内容如下：

```csharp
using UnityEngine;
using UnityEngine.UI;
using UnityEngine.EventSystems;
namespace Kanamori.Dbg
{
    public class DbgModelController : MonoBehaviour
    {
        private GameController game;
        public Transform frontCamera;
        public Transform ssMap;
        public GameObject blueBox;
        public GameObject addUI;
        public GameObject saveUI;
        private UIControlObject uiControl;
        public Text textShow;
        void Start()
        {
            game = FindObjectOfType<GameController>();
            uiControl = FindObjectOfType<UIControlObject>();
            addUI.SetActive(true);
            saveUI.SetActive(false);
            Load();
        }
        public void Back()
        {
            game.BackDbgMenu();
        }
        public void Add()
        {
            Instantiate(blueBox, frontCamera.position, Quaternion.identity, ssMap);
        }
```

```csharp
public void Close()
{
    addUI.SetActive(true);
    saveUI.SetActive(false);
}
void Update()
{
    if (Application.platform == RuntimePlatform.WindowsEditor)
    {
        if (Input.GetMouseButtonDown(0)
        && !EventSystem.current.IsPointerOverGameObject())
        {
            Ray ray = Camera.main.ScreenPointToRay(Input.mousePosition);
            TouchedObject(ray);
        }
    }
    else
    {
        if (Input.touchCount == 1
        && Input.touches[0].phase == TouchPhase.Began
        && !EventSystem.current.IsPointerOverGameObject(Input.touches[0].fingerId))
        {
            Ray ray = Camera.main.ScreenPointToRay(Input.touches[0].position);
            TouchedObject(ray);
        }
    }
}
private void TouchedObject(Ray ray)
{
    if (Physics.Raycast(ray, out RaycastHit hit))
    {
        addUI.SetActive(false);
        saveUI.SetActive(true);
        uiControl.SetSelected(hit.transform);
        textShow.text="选中物体";
    }
}
public void Save()
{
    string[] jsons = new string[ssMap.childCount];
    for (int i = 0; i < ssMap.childCount; i++)
    {
        DynamicObject dynamicObject = new DynamicObject();
        dynamicObject.position = ssMap.GetChild(i).localPosition;
        dynamicObject.rotation = ssMap.GetChild(i).localEulerAngles;
```

```csharp
            dynamicObject.scale = ssMap.GetChild(i).localScale;
            jsons[i] = JsonUtility.ToJson(dynamicObject);
        }
        game.SaveDynamicObject(jsons);
        textShow.text = "保存" + ssMap.childCount + "个游戏对象";
    }
    public void Delete()
    {
        var go = uiControl.selected.gameObject;
        uiControl.ClearSelected();
        Destroy(go);
        textShow.text = "删除选中物体,请保存结果。";
    }
    private void Load()
    {
        var list = game.LoadDynamicObject();
        foreach (var item in list)
        {
            var dynamicObject = JsonUtility.FromJson<DynamicObject>(item);
            var tf = Instantiate(blueBox, ssMap).transform;
            tf.localPosition = dynamicObject.position;
            tf.localEulerAngles = dynamicObject.rotation;
            tf.localScale = dynamicObject.scale;
        }
    }
}
```

8.4 关键点场景开发

8.4.1 场景搭建

1. 基础内容

这里也涉及界面切换,因为第一个界面只有返回按钮,所以就不再用其他游戏对象作为父级了。

打开 DbgKeyPoint 场景,单击菜单 GameObject→UI→Button,往场景中添加一个按钮。设置其 Anchor Presets 锚点预设为 bottom-stretch(底部对齐),设置其 Rect Transform 矩阵变换 "Left, Right, Pos Y, Height, Pos Z" 为 "0, 0, 70, 140, 0",修改按钮名称为 ButtonBack,如图 8-69 所示。

选中按钮的 Text 游戏对象,设置 Text 为 "返回",修改 Font 字体属性为导入字体,选中 Best Fit,修改 Max Size 属性为 100,如图 8-70 所示。

单击菜单 GameObject→UI→Panel,添加一个面板。按照前面章节内容在面板下添加按钮、文本和输入文本框,如图 8-71 所示。按钮名称文本和位置如表 8-7 所示。

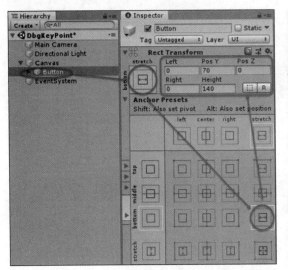
图 8-69

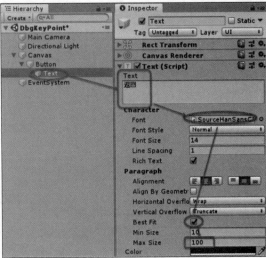
图 8-70

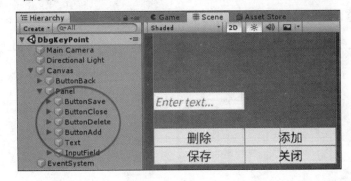
图 8-71

表 8-7 按钮名称文本和位置

类　型	Name（名称）	Text（文本）	Anchors（锚点）（Min X, Min Y, Max X, Max Y）	Rect Transform（矩阵变化）（Left, Right, Pos Y, Height, Pos Z）
按钮	ButtonSave	保存	0, 0, 0.5, 0	0, 0, 70, 140, 0
按钮	ButtonClose	关闭	0.5, 0, 1, 0	0, 0, 70, 140, 0
按钮	ButtonDelete	删除	0, 0, 0.5, 0	0, 0, 210, 140, 0
按钮	ButtonAdd	添加	0.5, 0, 1, 0	0.5, 0, 1, 0
文本	Text		0, 0, 1, 0	0, 0, 350, 140, 0
文本输入框	Inputfield		0, 0, 0.5, 0	0, 0, 490, 140, 0

2. 添加设置下拉列表

（1）添加下拉列表

选中 Panel 游戏对象，单击鼠标右键，在弹出的菜单中单击 UI→Dropdown，在 Panel 游戏对象下添加一个下拉列表，如图 8-72 所示。

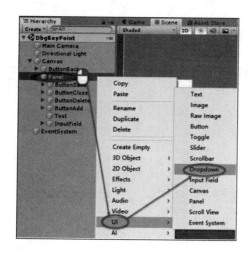

图 8-72

设置其 Anchor Presets 锚点预设为 bottom-stretch（底部对齐），设置其 Rect Transform 矩阵变换 "Left，Right，Pos Y，Height，Pos Z" 为 "0，0，490，140，0"，如图 8-73 所示。

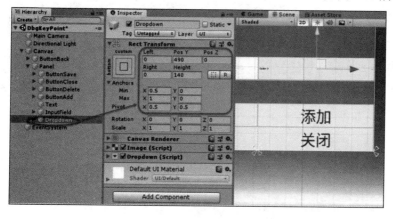

图 8-73

（2）设置未点击时候的样子

选中下拉列表的 Dropdown/Arrow 游戏对象，设置其 Rect Transform 矩阵变换 "Pos X，Width，Pos Y，Height，Pos Z" 为 "-70，140，0，140，0"。这样，右边的箭头就变大了，如图 8-74 所示。

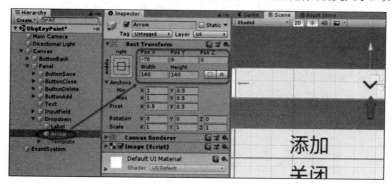

图 8-74

选中下拉列表的 Dropdown/Label 游戏对象，设置 Rect Transform 矩阵变换的 Right 为 145，设置其 Text 组件的 Font 字体为上传字体，选中 Best Fit，修改 Max Size 为 100，如图 8-75 所示。

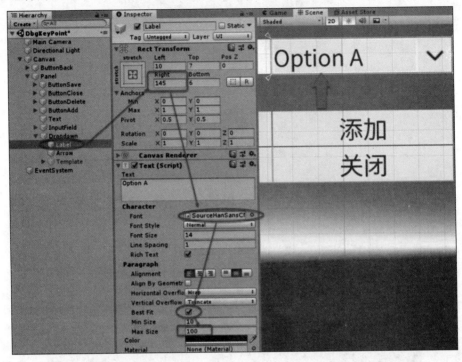

图 8-75

（3）设置下拉项目格式

选中 Dropdown/Template/Viewport/Content/Item/Item Label 游戏对象，设置其 Text 组件的 Font 字体为上传字体，选中 Best Fit，修改 Max Size 为 100，如图 8-76 所示。

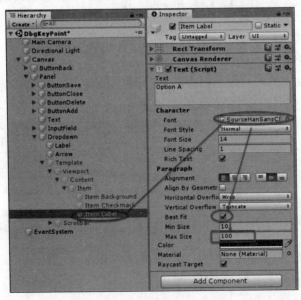

图 8-76

选中 Dropdown/Template/Viewport/Content/Item 游戏对象，设置 Rect Transform 的 Height 为 140，如图 8-77 所示。

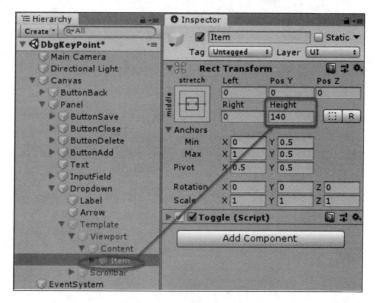

图 8-77

选中 Dropdown/Template/Viewport/Content 游戏对象，设置 Rect Transform 的 Height 为 148，如图 8-78 所示。

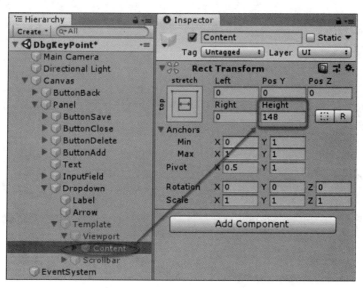

图 8-78

（4）设置下拉框大小

选中 Dropdown/Template 游戏对象，设置 Rect Transform 的 Height 为 500，如图 8-79 所示。此时下拉框的效果如图 8-80 所示。

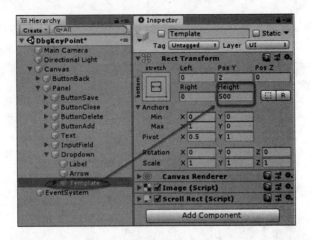

图 8-79

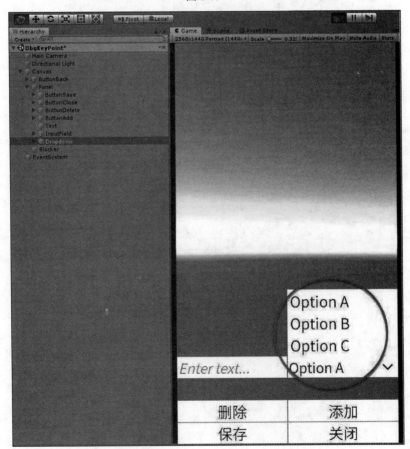

图 8-80

3. 添加设置滚动视图

（1）添加滚动视图

选中 Panel 游戏对象，单击鼠标右键，在弹出的菜单中选择 UI→Scroll View，添加一个滚动视图，如图 8-81 所示。

设置其 Anchor Presets 锚点预设为 stretch-stretch、其 Rect Transform 的"Left，Right，Top，Bottom，Pos Z"为"0，0，0，600，0"，即屏幕对齐、距离底部 600，如图 8-82 所示。

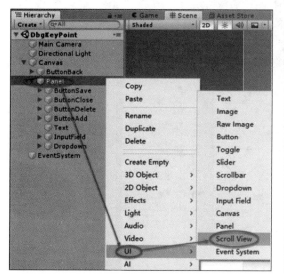
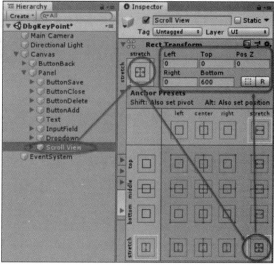

图 8-81　　　　　　　　　　　　　　　　图 8-82

（2）添加点击按钮

选中 Scroll View/Viewport/Content 游戏对象，单击鼠标右键，在弹出的菜单中选中 UI→Button，在滚动视图中添加一个按钮，如图 8-83 所示。

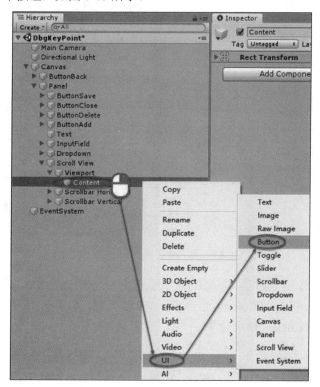

图 8-83

（3）设置滚动视图元素自动排布

选中 Scroll View 游戏对象，取消 Horizontal 选项（取消水平滚动条），如图 8-84 所示。

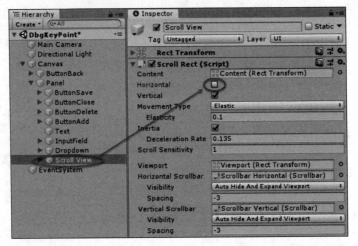

图 8-84

选中 Scroll View/Viewport/Content 游戏对象，单击 Add Component 按钮，选中 Layout 标签下的 Content Size Fitter（自动调整内容大小组件）和 Vertical Layout Group（垂直布局组件），如图 8-85 所示。

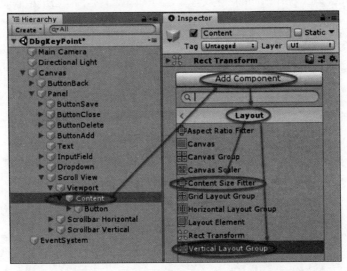

图 8-85

设置 Content Size Fitter 组件的 Vertical Fit 属性为 Min Size，设置 Vertical Layout Group 组件的 Spacing 间距为 3，并选中 Child Controls Size 下的 Width 选项，如图 8-86 所示。这样滚动视图的元素宽度会自动设置为和 Content 游戏对象一样宽，并垂直排列。

（4）设置按钮并生成预制件

选中按钮的 Text 游戏对象，设置 Font 字体为上传字体，选中 Best Fit，修改 Max Size 为 100，如图 8-87 所示。修改按钮名称为 ButtonSelected，如图 8-88 所示。

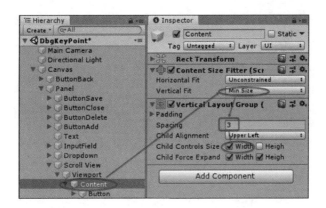

图 8-86

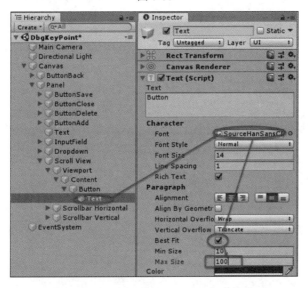

图 8-87

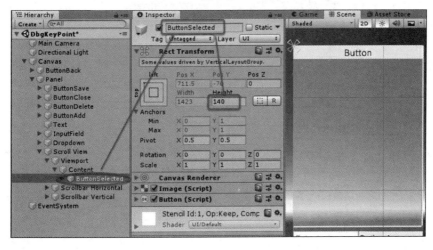

图 8-88

将按钮拖到 Kanamori/Prefabs 目录下生成预制件，并删除场景中的按钮，如图 8-89 所示。

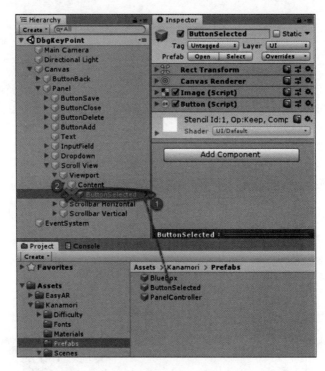

图 8-89

（5）添加其他游戏对象

添加一个空的游戏对象，命名为 FakeSSMap 当作稀疏空间地图的游戏对象。新建一个空的游戏对象，命名为 FakeImageTarget，当作平面图像跟踪的游戏对象，并在其下添加一个方块，如图 8-90 所示。

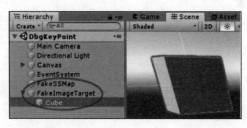

图 8-90

8.4.2 编写脚本

场景的主要逻辑（见图 8-91）有 4 种状态。开始是在等待选中物体的状态，当点击到方块以后，将点击的对象赋值给 Selected 属性记录，进入主界面的等待添加按钮状态。

（1）点击"添加"按钮，在滚动视图里添加一个按钮，并将选中对象信息赋值到按钮上。

（2）点击"保存"按钮，遍历滚动视图中的按钮，获取关键点信息并保存到文本文件。

（3）点击滚动视图中的按钮，会将按钮对象赋值给 Selected 属性记录，进入等待删除按钮状态。

（4）点击"删除"按钮，把滚动视图中对应的按钮删除。

无论是添加还是删除都是临时的，必须点击"保存"按钮后才有效果。

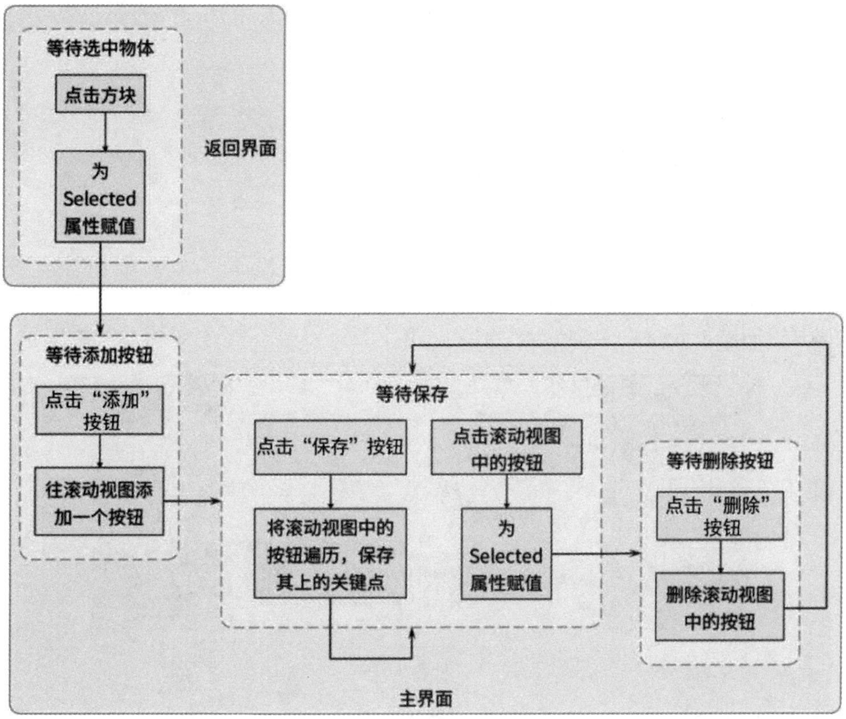

图 8-91

1. 添加脚本

添加一个空的游戏对象并命名为 SceneMaster。在目录 Kanamori/Scripts/Dbg 下添加脚本 DbgKeyPointController，并将其拖到 SceneMaster 游戏对象下成为其组件，如图 8-92 所示。

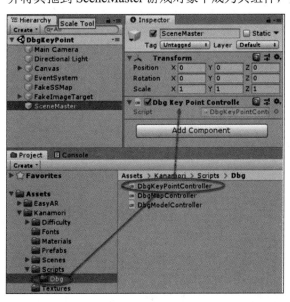

图 8-92

为了方便调试，需要移动方块，移动后希望摄像机永远都能看到方块。添加摄像机和图像的公开属性，使用 LookAt 方法让摄像机对准方块所在的游戏对象。

```
public Transform cam;
public Transform image;
void Update()
{
    cam.LookAt(image);
}
```

选中 SceneMaster 游戏对象，将 Main Camera 游戏对象拖到 Cam 属性中，将 Cube 游戏对象拖到 Image 属性中，如图 8-93 所示。这样，无论 Cube 游戏对象怎样移动都会出现在屏幕中间。

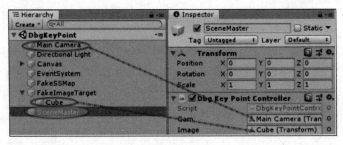

图 8-93

2. 返回和界面切换

（1）返回按钮方法

添加返回方法：

```
public void Back(){
    game.BackDbgMenu();
}
```

设置 ButtonBack 按钮的点击事件为 SceneMaster 游戏对象上 DbgKeyPointController 脚本下的 Back 方法，如图 8-94 所示。

（2）关闭按钮方法

添加界面显示的 2 个公开属性，添加关闭主要界面的方法，并在 Start 方法中调用。

```
public GameObject uiBack;
public GameObject uiMain;
void Start()
{
        Close();
}
public void Close()
{
        uiMain.SetActive(false);
        uiBack.SetActive(true);
}
```

第 8 章 调试场景开发

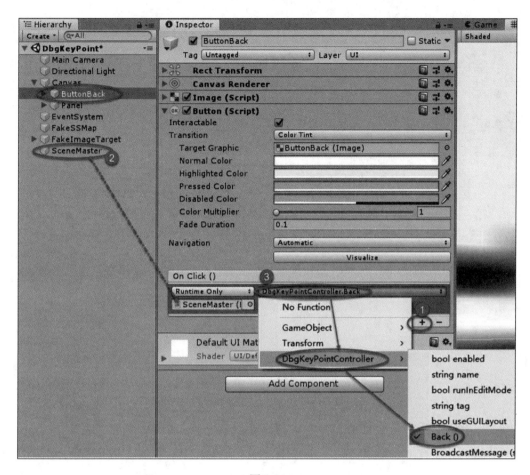

图 8-94

选中 SceneMaster 游戏对象，将 ButtonBack 游戏对象拖到 Ui Back 属性中，并为其赋值，将 Panel 游戏对象拖到 Ui Main 属性中，如图 8-95 所示。

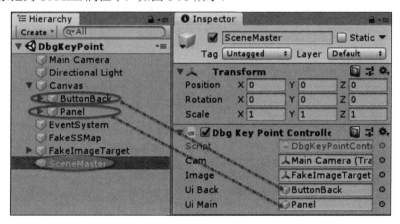

图 8-95

设置 ButtonClose 按钮的点击事件为 SceneMaster 游戏对象上 DbgKeyPointController 脚本下的 Close 方法，如图 8-96 所示。

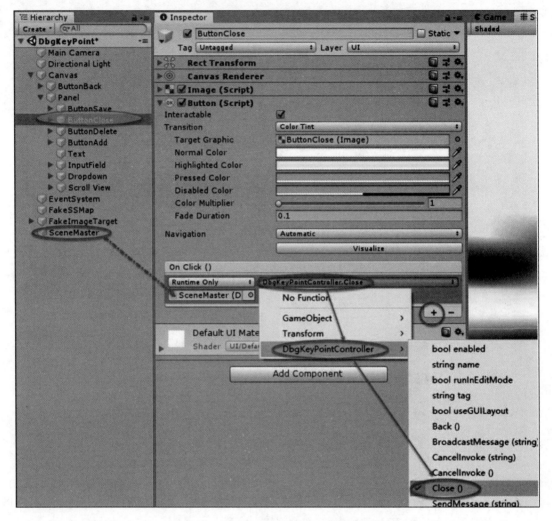

图 8-96

（3）点击后打开主界面的方法

和上一个场景一样需要对点击进行判断，发出射线。点击物体后，打开主界面。

```
void Update()
{
    cam.LookAt(image);
    if (Application.platform == RuntimePlatform.WindowsEditor)
    {
        if (Input.GetMouseButtonDown(0)
        && !EventSystem.current.IsPointerOverGameObject())
        {
            Ray ray = Camera.main.ScreenPointToRay(Input.mousePosition);
            TouchedObject(ray);
        }
    }
    else
```

```
        {
            if (Input.touchCount == 1
                && Input.touches[0].phase == TouchPhase.Began
                && !EventSystem.current.IsPointerOverGameObject (Input.touches[0].fingerId))
            {
                Ray ray = Camera.main.ScreenPointToRay(Input.touches[0].position);
                TouchedObject(ray);
            }
        }
    }
    private void TouchedObject(Ray ray)
    {
        if (Physics.Raycast(ray, out RaycastHit hit))
        {
            uiBack.SetActive(false);
            uiMain.SetActive(true);
        }
    }
```

3. 添加关键点类和保存方法

在 Kanamori/Scripts 目录下添加脚本 KeyPoint。其中，关键点类型用整数来区分，目的是方便下拉菜单的设置。为了避免后续编写代码的时候出错，一定要立即写上该属性的注释。

```
using UnityEngine;
using System;

namespace Kanamori
{
    [Serializable]
    public class KeyPoint
    {
        public Vector3 position;
        public string name;
        /// <summary>
        /// 类型：0=目的地；1=途经点
        /// </summary>
        public int pointType;
    }
}
```

选择 Dropdown 游戏对象，设置 Options 属性值，第 1 个是"目的地"，第 2 个是"途经点"，删除其他的选项，如图 8-97 所示。

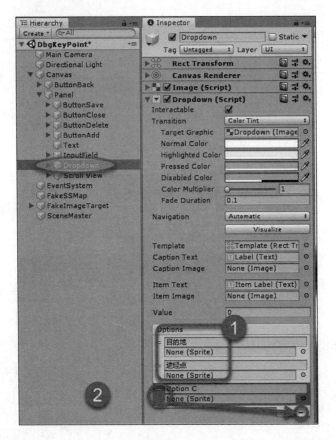

图 8-97

修改 GameController 脚本，添加关键点保存的路径和保存读取的方法。

```
private static readonly string pathKeyPoint = "/keypoint.txt";
public void SaveKeyPoint(string[] stringArray)
{
    SaveStringArray(stringArray, Application.persistentDataPath + pathKeyPoint);
}
public List<string> LoadKeyPoint()
{
    return LoadStringList(Application.persistentDataPath + pathKeyPoint);
}
```

4. 新增关键点

（1）修改 ButtonSelected Prefab 预制件

为了让滚动视图中的按钮能够记录关键点信息，需要修改预制件。

选中 Kanamori/Prefabs 目录下的 ButtonSelected Prefab 预制件，单击 Inspector 窗口的 Open Prefab 按钮，如图 8-98 所示。

在 Kanamori/Scripts 目录下新建脚本 SelectButton 并将其拖到 ButtonSelected 游戏对象下，然后关闭 Prefab 预制件，如图 8-99 所示。

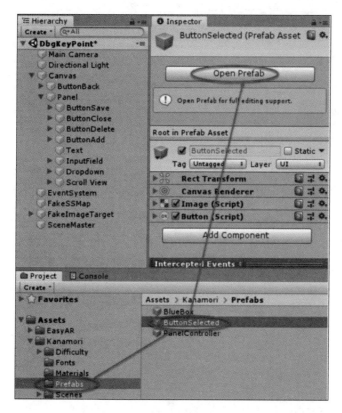

图 8-98

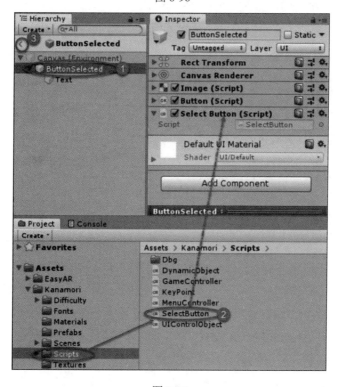

图 8-99

修改 SelectButton 脚本，添加一个关键点属性即可。

```csharp
using UnityEngine;
namespace Kanamori
{
    public class SelectButton : MonoBehaviour
    {
        public KeyPoint keyPoint;
    }
}
```

（2）添加 Add 方法

在 DbgKeyPointController 脚本中添加 Add 方法，用于给滚动视图添加按钮，并将 Selected 的位置信息赋值给按钮中 SelectButton 组件的 KeyPoint 属性。

添加判断和提示，为了避免重复添加，添加后要设置按钮不能点击。

```csharp
private Transform selected;
public Button btnAdd;
public Text textInfo;
public InputField inputField;
public Dropdown dropdown;
public Transform svContent;
public SelectButton prefab;
public Transform map;
void Start()
{
    ...
    btnAdd.interactable = false;
}
private void TouchedObject(Ray ray)
{
    if (Physics.Raycast(ray, out RaycastHit hit))
    {
        uiBack.SetActive(false);
        uiMain.SetActive(true);
        var tf = new GameObject().transform;
        tf.position = hit.transform.position;
        tf.parent = map.transform;
        selected = tf;
        btnAdd.interactable = true;
    }
}
public void Add()
{
    if (!string.IsNullOrEmpty(inputField.text) && selected != null)
    {
        SelectButton btn = Instantiate(prefab, svContent);
```

```
            btn.keyPoint.name = inputField.text;
            btn.keyPoint.position = selected.localPosition;
            btn.keyPoint.pointType = dropdown.value;

            btn.GetComponentInChildren<Text>().text = inputField.text;

            inputField.text = "";
            selected = null;
            textInfo.text = "添加成功。";
            btnAdd.interactable = false;
        }
    }
```

（3）设置脚本

选中 SceneMaster 游戏对象，将 ButtonSelected 预制件拖到 Prefab 属性中，如图 8-100 所示。游戏对象对应属性如表 8-8 所示。

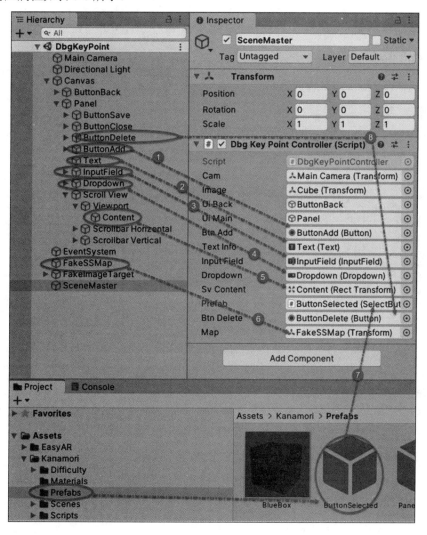

图 8-100

表 8-8　为脚本属性赋值

游戏对象	属　　性
ButtonAdd	Btn Add
Text	Text Info
InputField	Input Field
Dropdown	Dropdown
Content	Sv Content
FakeSSMap	Map
ButtonSelected（预制件）	Prefab
ButtonDelete	Btn Delete

设置 ButtonAdd 按钮的点击事件为 SceneMaster 游戏对象上 DbgKeyPointController 脚本下的 Add 方法，如图 8-101 所示。

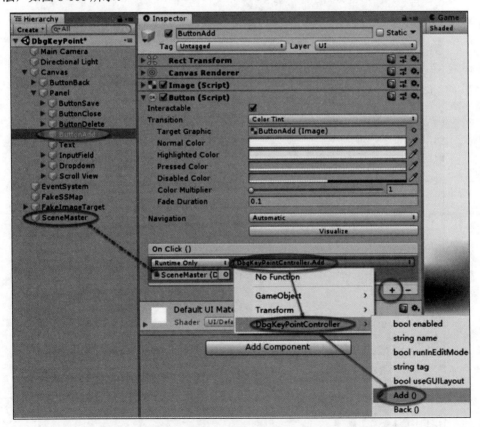

图 8-101

此时，从菜单场景开始运行，点击方块后，输入名称，就能点击"添加"按钮添加了。可以选中添加的按钮的游戏对象，查看添加内容是否正确，如图 8-102 所示。

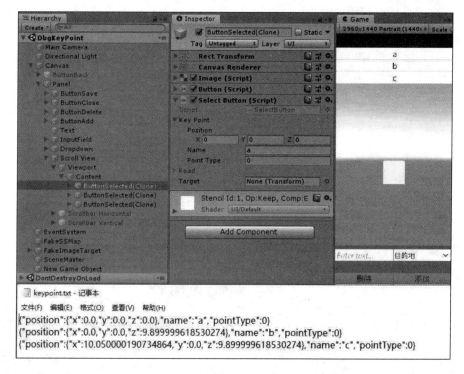

图 8-102

5. 保存关键点

添加保存关键点的方法，遍历滚动视图下的子游戏对象，获取 SelectButton 组件上的 KeyPoint 属性。添加加载的方法并在 Start 方法中调用。

```
void Start()
{
    game = FindObjectOfType<GameController>();
    Load ();
    Close();
    btnAdd.interactable = false;
}
public void Save()
{
    string[] jsons = new string[svContent.childCount];
    for (int i = 0; i < svContent.childCount; i++)
    {
        jsons[i] = JsonUtility.ToJson(
            svContent.GetChild(i).GetComponent<SelectButton>().keyPoint);
    }
    game.SaveKeyPoint(jsons);
    textInfo.text = "保存完成。";
}
private void Load ()
{
```

```
        var list = game.LoadKeyPoint();
        foreach (var item in list)
        {
            SelectButton btn = Instantiate(prefab, svContent);
            btn.keyPoint = JsonUtility.FromJson<KeyPoint>(item);
            btn.GetComponentInChildren<Text>().text = btn.keyPoint.name;
        }
    }
```

设置 ButtonSave 按钮的点击事件为 SceneMaster 游戏对象上 DbgKeyPointController 脚本下的 Save 方法，如图 8-103 所示。

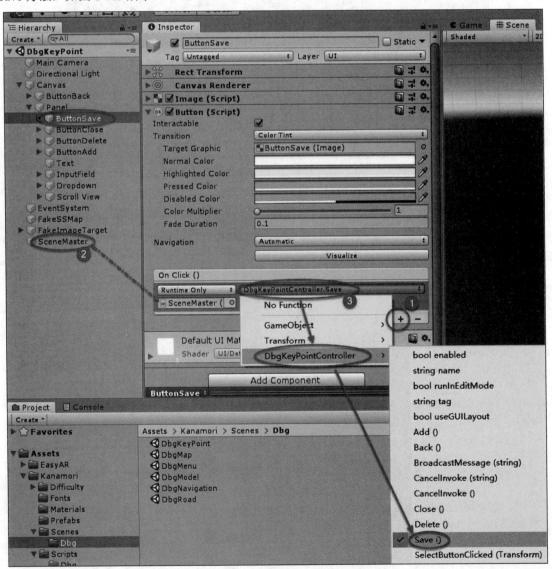

图 8-103

此时运行，就能将按钮保存到文本中，如图 8-104 所示。

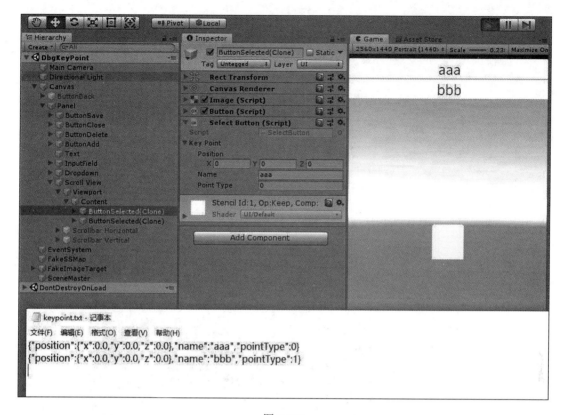

图 8-104

6. 删除关键点

当按钮被点击的时候,利用 SendMessage 方法把当前点击的按钮信息传给 DbgKeyPointController 脚本处理。

(1)修改 SelectButton 脚本

修改 SelectButton 脚本,在 Start 方法中获取当前游戏对象上的按钮组件,并添加点击事件对应的方法。用 SendMessage 方式调用 SceneMaster 游戏对象上的 SelectButtonClicked 方法。

```
void Start()
{
    gameObject.GetComponent<Button>().onClick.AddListener(() =>
    {
        GameObject.Find("SceneMaster")
            .SendMessage("SelectButtonClicked", transform);
    });
}
```

(2)添加响应方法

修改 DbgKeyPointController 脚本,添加 SelectButtonClicked 方法接收并处理对象。

```
public void SelectButtonClicked(Transform btn)
{
```

```
    selected = btn;
    textInfo.text = btn.GetComponentInChildren<Text>().text;
    btnDelete.interactable = true;
    btnAdd.interactable = false;
}
```

添加删除的方法：

```
public void Delete()
{
    Destroy(selected.gameObject);
    textInfo.text = "删除完成。";
    btnDelete.interactable = false;
}
```

设置 ButtonSave 按钮的点击事件为 SceneMaster 游戏对象上 DbgKeyPointController 脚本下的 Save 方法，如图 8-105 所示。

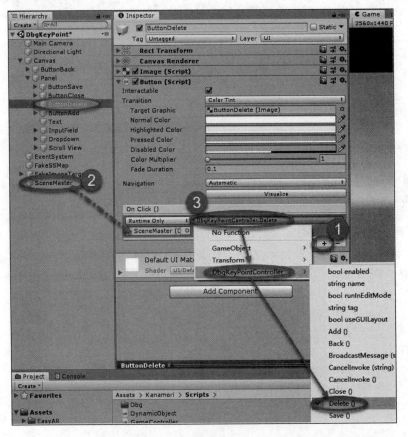

图 8-105

此时，该场景功能完成。添加几个关键点为后面做准备，如图 8-106 所示。

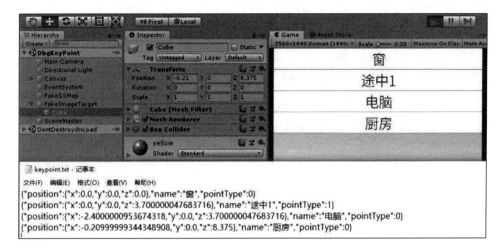

图 8-106

完整脚本内容如下：

```csharp
using UnityEngine;
using UnityEngine.EventSystems;
using UnityEngine.UI;

namespace Kanamori.Dbg
{
    public class DbgKeyPointController : MonoBehaviour
    {
        public Transform cam;
        public Transform image;
        private GameController game;
        public GameObject uiBack;
        public GameObject uiMain;
        private Transform selected;
        public Button btnAdd;
        public Text textInfo;
        public InputField inputField;
        public Dropdown dropdown;
        public Transform svContent;
        public SelectButton prefab;
        public Button btnDelete;
        public Transform map;
        void Start()
        {
            game = FindObjectOfType<GameController>();
            Load();
            btnAdd.interactable = false;
            btnDelete.interactable = false;
            Close();
        }
```

```csharp
void Update()
{
    cam.LookAt(image);
    if (Application.platform == RuntimePlatform.WindowsEditor)
    {
        if (Input.GetMouseButtonDown(0)
        && !EventSystem.current.IsPointerOverGameObject())
        {
            Ray ray = Camera.main.ScreenPointToRay(Input.mousePosition);
            TouchedObject(ray);
        }
    }
    else
    {
        if (Input.touchCount == 1
        && Input.touches[0].phase == TouchPhase.Began
        &&!EventSystem.current.IsPointerOverGameObject(Input.touches[0].fingerId))
        {
            Ray ray = Camera.main.ScreenPointToRay(Input.touches[0].position);
            TouchedObject(ray);
        }
    }
}
private void TouchedObject(Ray ray)
{
    if (Physics.Raycast(ray, out RaycastHit hit))
    {
        uiBack.SetActive(false);
        uiMain.SetActive(true);
        var tf = new GameObject().transform;
        tf.position = hit.transform.position;
        tf.parent = map.transform;
        selected = tf;
        btnAdd.interactable = true;
    }
}
public void Back()
{
    if (game)
    {
        game.BackDbgMenu();
    }
}
public void Close()
{
```

```csharp
            uiMain.SetActive(false);
            uiBack.SetActive(true);
        }
        public void Add()
        {
            if (!string.IsNullOrEmpty(inputField.text) && selected != null)
            {
                SelectButton btn = Instantiate(prefab, svContent);

                btn.keyPoint.name = inputField.text;
                btn.keyPoint.position = selected.localPosition;
                btn.keyPoint.pointType = dropdown.value;

                btn.GetComponentInChildren<Text>().text = inputField.text;

                inputField.text = "";
                selected = null;
                textInfo.text = "添加完成。";
                btnAdd.interactable = false;
            }
        }
        public void Save()
        {
            string[] jsons = new string[svContent.childCount];
            for (int i = 0; i < svContent.childCount; i++)
            {
                jsons[i] = JsonUtility.ToJson(svContent.GetChild(i).GetComponent<SelectButton>().keyPoint);
            }
            if (game)
            {
                game.SaveKeyPoint(jsons);
                textInfo.text = "保存完成。";
            }
        }
        private void Load()
        {
            if (game)
            {
                var list = game.LoadKeyPoint();
                foreach (var item in list)
                {
                    SelectButton btn = Instantiate(prefab, svContent);
                    btn.keyPoint = JsonUtility.FromJson<KeyPoint>(item);
                    btn.GetComponentInChildren<Text>().text = btn.keyPoint.name;
                }
            }
```

```
        }
        public void SelectButtonClicked(Transform btn)
        {
            selected = btn;
            textInfo.text = btn.GetComponentInChildren<Text>().text;
            btnDelete.interactable = true;
            btnAdd.interactable = false;
        }
        public void Delete()
        {
            Destroy(selected.gameObject);
            textInfo.text = "删除完成。";
            btnDelete.interactable = false;
        }
    }
}
```

8.5 预备路径场景开发

路径场景的界面和程序逻辑与关键点场景很接近，很多内容都可以通过复制粘贴，做简单修改即可完成。

8.5.1 场景设置

1. 复制界面

打开 DbgKeyPoint 场景，选中 Panel 游戏对象下的 UI，单击鼠标右键，在弹出的菜单中选择 Copy 复制游戏对象，如图 8-107 所示。

打开 DbgRoad 场景，单击菜单 GameObject→UI→Canvas，在场景中添加一个画布。将复制的 UI 贴到场景中并拖到 Canvas 游戏对象，如图 8-108 所示。

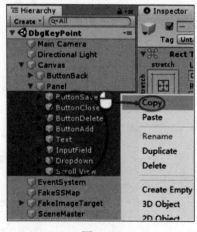

图 8-107

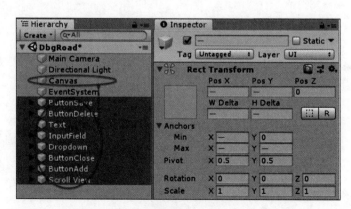

图 8-108

2. 修改界面

复制 Dropdown 游戏对象，设置 Anchors（Min X，Min Y，Max X，Max Y）为"0，0，0.5，0"，Rect Transform（Left，Right，Pos Y，Height，Pos Z）为"0，0，490，140，0"，修改名称为 dpdStart，如图 8-109 所示。

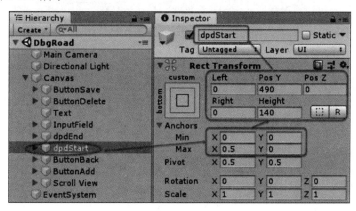

图 8-109

删除 dpdStart 下拉列表的 Options，如图 8-110 所示。修改原有下拉列表名称为 dpdEnd，删除下拉列表的 Options，如图 8-111 所示。

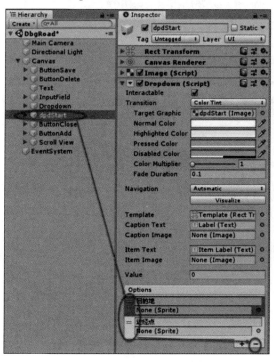 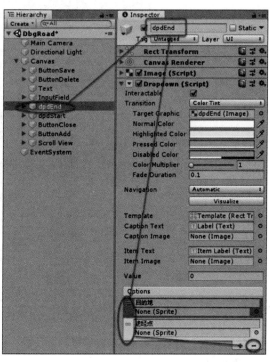

图 8-110　　　　　　　　　　　　图 8-111

修改 ButtonClose 游戏对象名称为 ButtonBack，修改显示文本为"返回"，如图 8-112 所示。

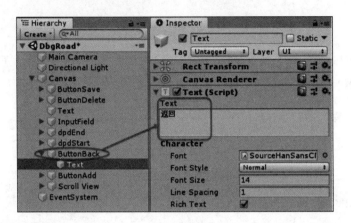

图 8-112

8.5.2 编写脚本

1. 添加脚本和返回

路径场景不需要用到稀疏空间地图,所以这个场景的脚本调试和正式使用的是同一个。

新建一个空的游戏对象并命名为 SceneMaster,在目录 Kanamori/Scripts 中新建脚本 RoadController,并将其拖到 SceneMaster 游戏对象中成为其组件,如图 8-113 所示。

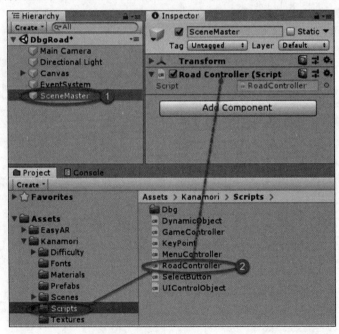

图 8-113

添加返回功能:

```
private GameController game;
void Start()
{
```

```
        game = FindObjectOfType<GameController>();
    }
    public void BackDbgMenu()
    {
        game.BackDbgMenu();
    }
```

设置 ButtonBack 按钮的点击事件为 SceneMaster 游戏对象上的 RoadController 脚本下的 BackDbgMenu 方法，如图 8-114 所示。

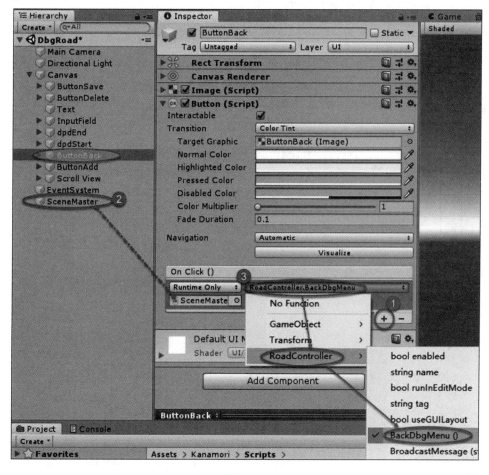

图 8-114

2. 添加路径类和保存方法

在目录 Kanamori/Scripts 中新建脚本，并命名为 Road 路径脚本，只需要公开属性。

```
using UnityEngine;
using System;

namespace Kanamori
{
    [Serializable]
```

```
public class Road
{
    public string startName;
    public Vector3 startPosition;
    public string endName;
    public Vector3 endPosition;
}
```

修改 GameController 脚本,添加路径存储的地址和对应方法:

```
private static readonly string pathRoad = "/road.txt";
public void SaveRoad(string[] stringArray)
{
    SaveStringArray(stringArray, Application.persistentDataPath + pathRoad);
}
public List<string> LoadRoad()
{
    return LoadStringList(Application.persistentDataPath + pathRoad);
}
```

3. 加载关键点填充下拉列表

加载关键点,绑定下拉列表:

```
public Dropdown dpdStart;
public Dropdown dpdEnd;
void Start()
{
    game = FindObjectOfType<GameController>();
    BindDropdown();
}
private void BindDropdown()
{
    var list = game.LoadKeyPoint();
    foreach (var item in list)
    {
        KeyPoint keyPoint = JsonUtility.FromJson<KeyPoint>(item);
        dpdStart.options.Add(new Dropdown.OptionData(keyPoint.name));
        dpdEnd.options.Add(new Dropdown.OptionData(keyPoint.name));
        dpdStart.captionText.text = dpdStart.options[0].text;
        dpdEnd.captionText.text = dpdEnd.options[0].text;
    }
}
```

选中 SceneMaster 游戏对象,将 dpdStart 游戏对象拖到 Dpd Start 属性中,将 dpdEnd 游戏对象拖到 Dpd End 属性中,如图 8-115 所示。

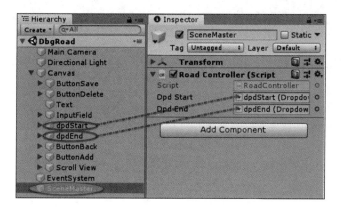

图 8-115

4. 添加路径

（1）添加按钮的路径属性

修改 SelectButton 脚本，添加一个路径属性。

```
public class SelectButton : MonoBehaviour
{
    public KeyPoint keyPoint;
    public Road road;
    ...
}
```

（2）添加 Add 方法

在 RoadController 脚本中添加 Add 方法，用于给滚动视图添加按钮，并将 Selected 的位置信息赋值给按钮的 SelectButton 组件的 road 属性。

```
public Text textInfo;
public Transform svContent;
public SelectButton prefab;
private List<KeyPoint> keyPoints;
void Start()
{
    ...
    btnAdd.interactable = false;
    keyPoints = new List<KeyPoint>();
}
public void Add()
{
    SelectButton btn = Instantiate(prefab, svContent);
    btn.road.startName = dpdStart.captionText.text;
    btn.road.endName = dpdEnd.captionText.text;
    btn.road.startPosition = GetPositionByName(btn.road.startName);
    btn.road.endPosition = GetPositionByName(btn.road.endName);
    btn.GetComponentInChildren<Text>().text = btn.road.startName + "<===>" + btn.road.endName;
```

```
            textInfo.text = "添加完成。";
    }
    private Vector3 GetPositionByName(string pName)
    {
        foreach (var kp in keyPoints)
        {
            if (kp.name == pName)
            {
                return kp.position;
            }
        }
        return Vector3.zero;
    }
```

(3) 设置脚本

选中 SceneMaster 游戏对象，将 Text 游戏对象拖到 Text Info 属性中，将 Content 游戏对象拖到 Sv Content 属性中，将 ButtonSelected 预制件拖到 Prefab 属性中，如图 8-116 所示。

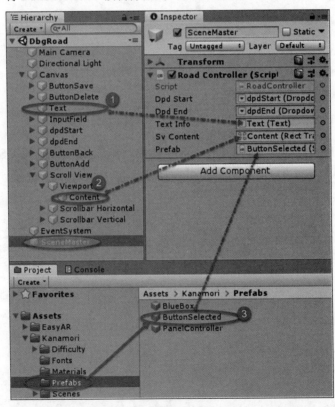

图 8-116

设置 ButtonAdd 按钮的点击事件为 SceneMaster 游戏对象上 RoadController 脚本下的 Add 方法，如图 8-117 所示。

此时，从菜单场景开始运行，选择下拉列表，单击"添加"按钮就能添加路径，如图 8-118 所示。可以选中添加按钮的游戏对象，查看添加的内容是否正确。

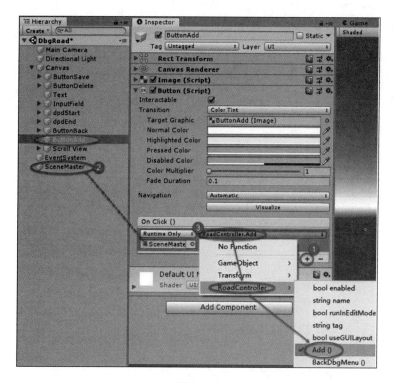

图 8-117

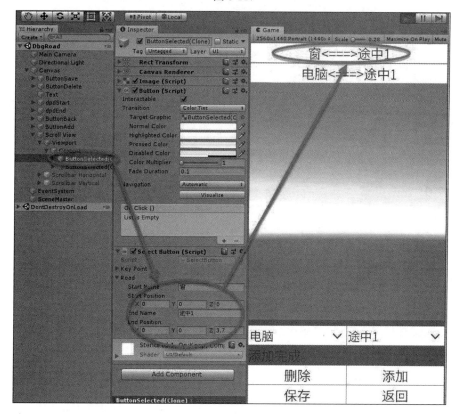

图 8-118

5. 保存删除路径

（1）添加保存方法

修改 RoadController 脚本，添加保存方法。遍历滚动视图的元素，获取按钮上的 road 属性，转换成 JSON 字符串后保存。

```
public void Save()
{
    string[] jsons = new string[svContent.childCount];
    for (int i = 0; i < svContent.childCount; i++)
    {
        jsons[i] = JsonUtility
.ToJson(svContent.GetChild(i).GetComponent<SelectButton>().road);
    }
    game.SaveRoad(jsons);
    textInfo.text = "保存完成。";
}
```

添加加载方法。从文本文件加载路径，生成按钮并赋值。

```
void Start()
{
    ...
    Load();
}
private void Load()
{
    var list = game.LoadRoad();
    foreach (var item in list)
    {
        var btn = Instantiate(prefab, svContent);
        btn.road = JsonUtility.FromJson<Road>(item);
        btn.GetComponentInChildren<Text>().text =
btn.road.startName + "<===>" + btn.road.endName;
    }
}
```

（2）添加删除方法

添加一个 selected 属性以获取当前点击的按钮，用同样的方法接收按钮点击发送过来的变量。

```
private Transform selected;
public Button btnDelete;
void Start()
{
    ...
    btnDelete.interactable = false;
}
public void SelectButtonClicked(Transform btn)
```

```
    {
        selected = btn;
        textInfo.text = btn.GetComponentInChildren<Text>().text;
        btnDelete.interactable = true;
    }
    public void Delete()
    {
        Destroy(selected.gameObject);
        textInfo.text = "删除完成。";
        btnDelete.interactable = false;
    }
```

选中 SceneMaster 游戏对象，将 ButtonDelete 拖到 Btn Delete 属性中，如图 8-119 所示。

设置 ButtonSave 按钮的点击事件为 SceneMaster 游戏对象上 RoadController 脚本下的 Save 方法。设置 ButtonDelete 按钮的点击事件为 SceneMaster 游戏对象上 RoadController 脚本下的 Delete 方法，如图 8-120 所示。

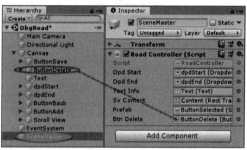

图 8-119

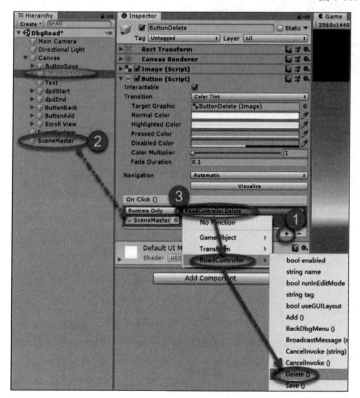

图 8-120

从菜单场景开始运行，可以看到功能都完成了。最后再添加路径供导航场景使用，如图 8-121 所示。

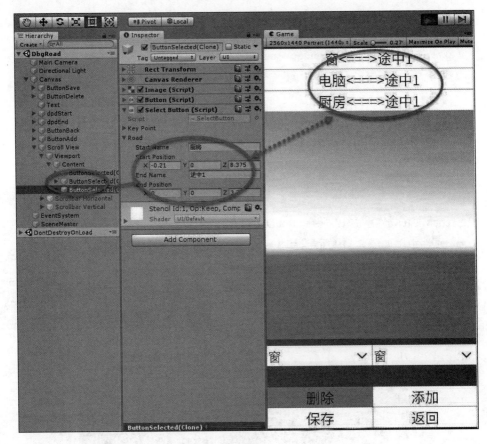

图 8-121

完整脚本内容如下：

```
using System.Collections.Generic;
using UnityEngine;
using UnityEngine.UI;

namespace Kanamori
{
    public class RoadController : MonoBehaviour
    {
        private GameController game;
        public Dropdown dpdStart;
        public Dropdown dpdEnd;
        public Text textInfo;
        public Transform svContent;
        public SelectButton prefab;
        private List<KeyPoint> keyPoints;
        private Transform selected;
        public Button btnDelete;

        void Start()
```

```csharp
    {
        game = FindObjectOfType<GameController>();
        keyPoints = new List<KeyPoint>();
        BindDropdown();
        btnDelete.interactable = false;
        Load();
    }
    public void BackDbgMenu()
    {
        if (game)
        {
            game.BackDbgMenu();
        }
    }
    public void BackMenu()
    {
        if (game)
        {
            game.BackMenu();
        }
    }
    private void BindDropdown()
    {
        if (game)
        {
            var list = game.LoadKeyPoint();
            foreach (var item in list)
            {
                KeyPoint keyPoint = JsonUtility.FromJson<KeyPoint>(item);
                dpdStart.options.Add(new Dropdown.OptionData(keyPoint.name));
                dpdEnd.options.Add(new Dropdown.OptionData(keyPoint.name));
                dpdStart.captionText.text = dpdStart.options[0].text;
                dpdEnd.captionText.text = dpdEnd.options[0].text;
                keyPoints.Add(keyPoint);
            }
        }
    }
    public void Add()
    {
        SelectButton btn = Instantiate(prefab, svContent);
        btn.road.startName = dpdStart.captionText.text;
        btn.road.endName = dpdEnd.captionText.text;
        btn.road.startPosition = GetPositionByName(btn.road.startName);
        btn.road.endPosition = GetPositionByName(btn.road.endName);
        btn.GetComponentInChildren<Text>().text = btn.road.startName + "<===>"
+ btn.road.endName;
        textInfo.text = "添加完成。";
```

```csharp
        }
        private Vector3 GetPositionByName(string pName)
        {
            foreach (var kp in keyPoints)
            {
                if (kp.name == pName)
                {
                    return kp.position;
                }
            }
            return Vector3.zero;
        }
        public void Save()
        {
            string[] jsons = new string[svContent.childCount];
            for (int i = 0; i < svContent.childCount; i++)
            {
                jsons[i] = JsonUtility.ToJson(svContent.GetChild(i).GetComponent<SelectButton>().road);
            }
            if (game)
            {
                game.SaveRoad(jsons);
                textInfo.text = "保存完成。";
            }
        }
        public void SelectButtonClicked(Transform btn)
        {
            selected = btn;
            textInfo.text = btn.GetComponentInChildren<Text>().text;
            btnDelete.interactable = true;
        }
        public void Delete()
        {
            Destroy(selected.gameObject);
            textInfo.text = "删除完成。";
            btnDelete.interactable = false;
        }
        private void Load()
        {
            if (game)
            {
                var list = game.LoadRoad();
                foreach (var item in list)
                {
                    var btn = Instantiate(prefab, svContent);
                    btn.road = JsonUtility.FromJson<Road>(item);
```

```
                    btn.GetComponentInChildren<Text>().text = btn.road.startName +
"<===>" + btn.road.endName;
                }
            }
        }
    }
}
```

8.6 导航场景开发

8.6.1 场景搭建

1. 添加切换用的游戏对象

单击菜单 GameObject→UI→Canvas，在场景里添加一个画布。在 Canvas 游戏对象下添加 2 个空的游戏对象，设置 Anchor Presets 为 stretch –stretch、Rect Transform 的"Left，Right，Top，Bottom，Pos Z"为"0，0，0，0，0"，如图 8-122 所示。修改游戏对象名称，一个命名为 UIBack，用于返回界面；一个命名为 UINav，用于导航界面。

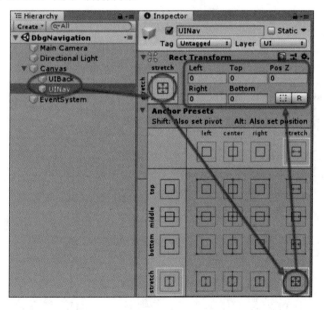

图 8-122

2. 添加返回界面

在 DbgMap 场景中，把画布 Canvas 游戏对象下的 UI 复制粘贴到 DbgNavigation 场景中，并拖动到 UIBack 游戏对象下。

修改 ButtonSave 游戏对象的名称为 ButtonNav，修改按钮文本显示为导航，如图 8-123 所示。

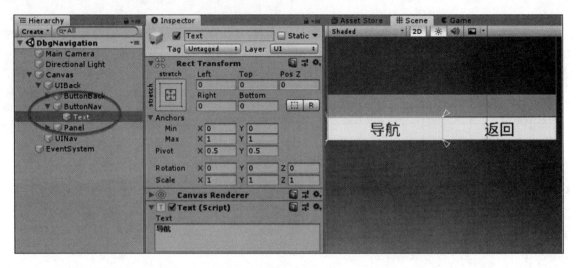

图 8-123

3. 添加导航界面

在 DbgRoad 场景中，把画布 Canvas 游戏对象下的 UI 复制粘贴到 DbgNavigation 场景中，并拖动到 UINav 游戏对象下。

保留 ButtonBack 和 Scroll View，删除其他游戏对象；修改 ButtonBack 游戏对象的名称为 ButtonClose，设置其 Anchor Presets 为 bottom-stretch、Rect Transform 的"Left，Right，Pos Y，Height，Pos Z"为"0，0，70，140，0"，如图 8-124 所示。修改按钮显示文本为"关闭"。

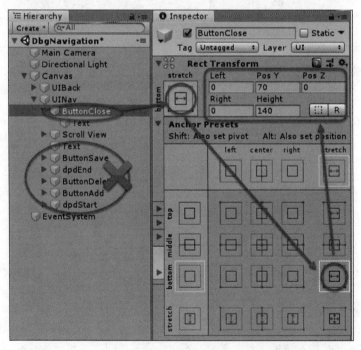

图 8-124

选中 Scroll View 游戏对象，设置其 Bottom 属性为 150，如图 8-125 所示。

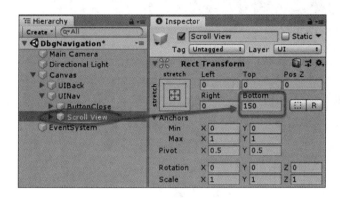

图 8-125

4. 其他

添加一个空的游戏对象，命名为 FakeSSMap，用于充当稀疏空间地图，如图 8-126 所示。

图 8-126

新建一个空的游戏对象并命名为 SceneMaster，在目录 Kanamori/Scripts/Dbg 中新建脚本 DbgNavigationController，并将其拖到 SceneMaster 游戏对象中成为其组件，如图 8-127 所示。

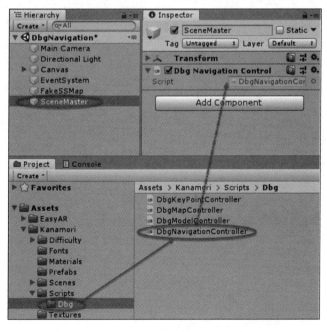

图 8-127

8.6.2 界面切换和返回

添加界面切换的公开属性和方法：

```
public GameObject uiBack;
public GameObject uiNav;
void Start()
{
    ShowNav();
}
public void Close()
{
    uiBack.SetActive(true);
    uiNav.SetActive(false);
}
public void ShowNav()
{
    uiBack.SetActive(false);
    uiNav.SetActive(true);
}
```

选中 SceneMaster 游戏对象，将 UIBack 游戏对象拖到 Ui Back 属性中，将 UINav 游戏对象拖到 Ui Nav 属性中，如图 8-128 所示。

图 8-128

添加游戏控制属性和返回方法：

```
private GameController game;
void Start()
{
    game = FindObjectOfType<GameController>();
    ShowNav();
}
public void Back()
{
    game.BackDbgMenu();
}
```

设置 ButtonBack 按钮的点击事件为 SceneMaster 游戏对象上 DbgNavigationController 脚本下的 Back 方法。设置 ButtonNav 按钮的点击事件为 SceneMaster 游戏对象上 DbgNavigationController 脚本下的 ShowNav 方法。设置 ButtonClose 按钮的点击事件为 SceneMaster 游戏对象上 DbgNavigationController 脚本下的 Close 方法，如图 8-129 所示。

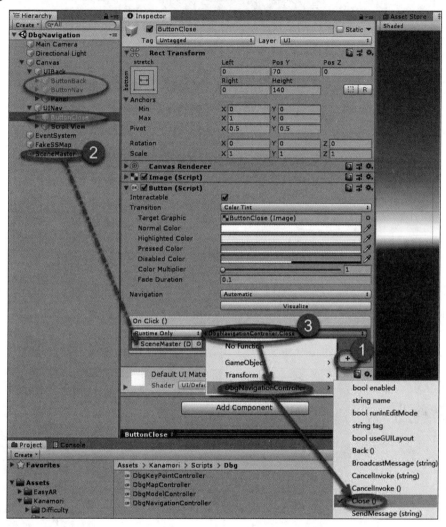

图 8-129

8.6.3 修改显示模型

这些模型不需要一直显示，只在使用者距离模型足够近的时候显示，离远以后就隐藏。这使用物理碰撞的穿透事件来实现。

1. 修改静态模型

新建一个空的场景，单击菜单 GameObject→3D Object→Sphere，添加一个球体。将 RPG Monster Duo PBR Polyart/Prefabs 目录下的 SlimePBR 预制件拖到 Sphere 游戏对象下，如图 8-130 所示。

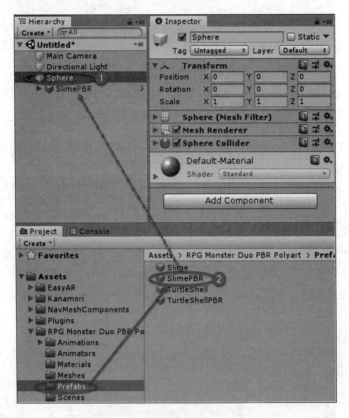

图 8-130

修改 SlimePBR 游戏对象的相对高度，使其基本在球体范围内，如图 8-131 所示。

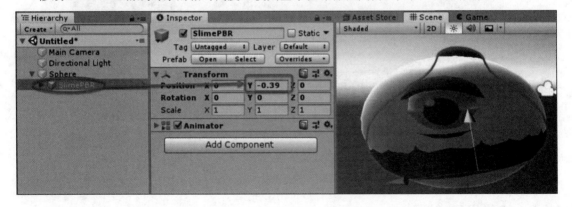

图 8-131

2. 设置显示并添加脚本

选中 Sphere 游戏对象，修改名称为 SlimeNew。取消 Mesh Renderer 组件的选中状态，这样就看不到了。在 Kanamori/Scripts 目录下新建脚本 ShowChildObject，并拖到 SlimeNew 游戏对象下成为其组件，如图 8-132 所示。

将 SlimeNew 游戏对象拖到 Kanamori/Prefabs 目录中成为预制件，如图 8-133 所示。

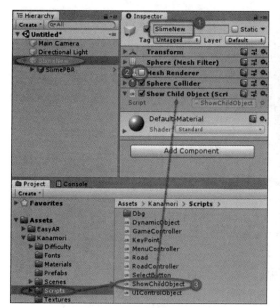

图 8-132

图 8-133

编写 ShowChildObject 脚本，添加穿透事件和将其子元素隐藏及显示的方法。

```
using UnityEngine;
namespace Kanamori
{
    public class ShowChildObject : MonoBehaviour
    {
        private void OnTriggerEnter(Collider other)
        {
            SetVisible(true);
        }
        private void OnTriggerExit(Collider other)
        {
            SetVisible(false);
        }
        public void SetVisible(bool status)
        {
            for (int i = 0; i < transform.childCount; i++)
            {
                transform.GetChild(i).gameObject.SetActive(status);
            }
        }
    }
}
```

3. 修改动态添加的盒子的预制件

选中 Kanamori/Prefabs 目录下的 BlueBox 预制件，单击 Open Prefab 按钮，打开预制件编辑，如图 8-134 所示。

在 Kanamori/Scripts 目录下新建脚本 ShowSelfObject，并将其拖到 BlueBox 游戏对象下成为其组件，点击 "<" 按钮完成编辑，如图 8-135 所示。

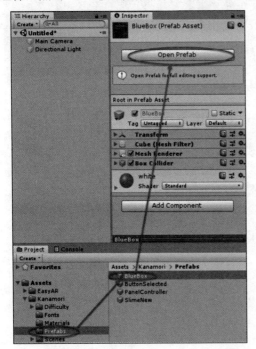
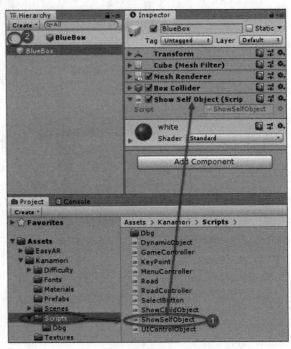

图 8-134　　　　　　　　　　　　　　　图 8-135

编写 ShowSelfObject 脚本，事件和方法与 ShowChildObejct 基本一致，区别只是通过自身的 MeshRenderer 组件是否激活来实现显示和隐藏效果。

```
using UnityEngine;
namespace Kanamori
{
    public class ShowSelfObject : MonoBehaviour
    {
        private void OnTriggerEnter(Collider other)
        {
            SetVisible(true);
        }
        private void OnTriggerExit(Collider other)
        {
            SetVisible(false);
        }
        public void SetVisible(bool status)
        {
            GetComponent<MeshRenderer>().enabled = status;
        }
    }
}
```

8.6.4 添加静态模型

添加静态模型其实只需要将模型放到稀疏空间地图的游戏对象下即可。具体位置和稀疏空间的位置有关，在这个阶段就不处理了，只是示意一下，这里主要处理靠近显示的问题。

1. 添加静态模型

选中 Kanamori/Prefabs 目录下的 SlimeNew，并拖到 FakeSSMap 游戏对象下，修改其位置和角度，如图 8-136 所示。

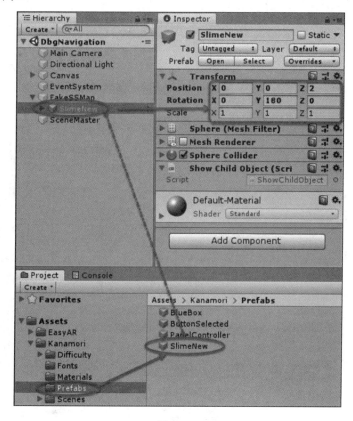

图 8-136

2. 修改摄像机设置

这里的摄像机位置就代表了使用者所在的位置。

选中 Main Camera 游戏对象，单击 Add Component→Physics，选中 Capsule Collider（胶囊碰撞器）和 Rigidbody（刚体），如图 8-137 所示。

选中 Capsule Collider 组件的 Is Trigger 选项，允许穿透；修改 Radius（半径）属性为 2，即靠近到模型 2 米的时候就会显示模型。去掉 Rigidbody 组件的 Use Gravity，就不会因为没有阻挡一直往下掉了，如图 8-138 所示。

此时，从菜单场景开始运行，当摄像头距离模型近的时候，模型就会显示，距离远的时候模型消失，如图 8-139 所示。

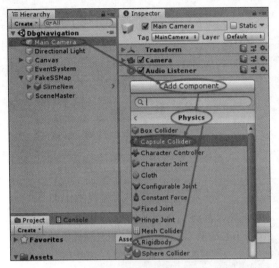
图 8-137

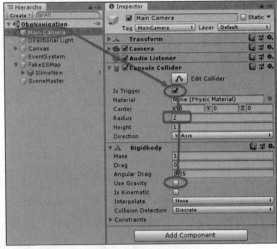
图 8-138

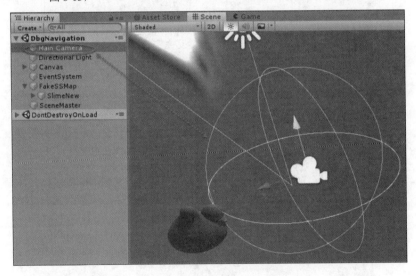
图 8-139

3. 修改 DbgNavigationController 脚本

添加用户属性和静态模型组的属性,在 Start 方法中判断用户和模型的直接距离,判断是否需要显示。

```
public Transform player;
public ShowChildObject[] staticObjects;
void Start()
{
    ...
    SetStaticObject();
}
private void SetStaticObject()
{
```

```
        foreach (var item in staticObjects)
        {
            item.SetVisible((item.transform.position - player.position).
magnitude<=2);
        }
    }
```

选中 SceneMaster 游戏对象，将 Main Camera 游戏对象拖到 Player 属性中并为其赋值，设置 Static Objects 属性的 Size 为 1，将 SlimeNew 拖到其元素属性中，如图 8-140 所示。这样，开始的时候就能根据距离自动显示或隐藏。

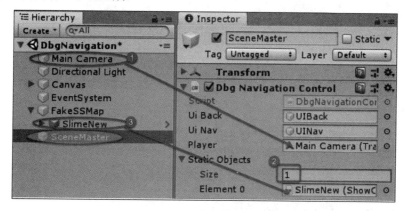

图 8-140

8.6.5 添加模型场景对应模型

这里的脚本和模型场景的脚本很接近，只是多了判断距离后是否显示的内容。

```
public Transform ssMap;
public Transform blueBox;
void Start()
{
    ...
    LoadObjects();
    ShowNav();
}
public void LoadObjects()
{
    var list = game.LoadDynamicObject();
    foreach (var item in list)
    {
        var dynamicObject = JsonUtility.FromJson<DynamicObject>(item);
        var tf = Instantiate(blueBox, ssMap);
        tf.localPosition = dynamicObject.position;
        tf.localEulerAngles = dynamicObject.rotation;
        tf.localScale = dynamicObject.scale;
        var obj = tf.GetComponent<ShowSelfObject>();
```

```
            obj.SetVisible((tf.position - player.position).magnitude <= 2);
        }
    }
```

选中 SceneMaster 游戏对象,将 FakeSSMap 拖到 Ss Map 属性中,将 Kanamori/Prefabs 目录下的 BlueBox 预制件拖到 Blue Box 属性中,如图 8-141 所示。

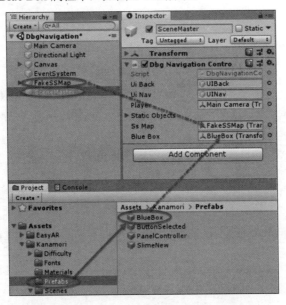

图 8-141

从菜单场景运行,进入导航场景以后会自动添加蓝色盒子,当摄像机接近时盒子会自动显示,如图 8-142 所示。

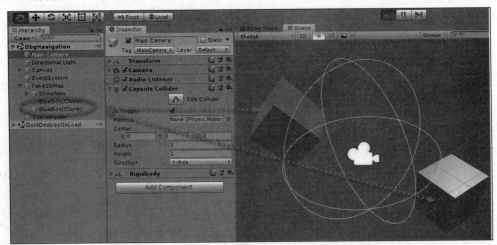

图 8-142

8.6.6 添加关键点

这里的关键点会变成目的地按钮,变成到终点显示的模型。

1. 添加终点模型预制件

新建一个场景,在场景中添加一个空的游戏对象并命名为 TurtleShellNew,将 RPG Monster Duo PBR Polyart/Prefabs 目录下的 TurtlesShellPBR 预制件拖到 TurtleShellNew 游戏对象下,并修改 TurtlesShellPBR 游戏对象的位置和大小,如图 8-143 所示。

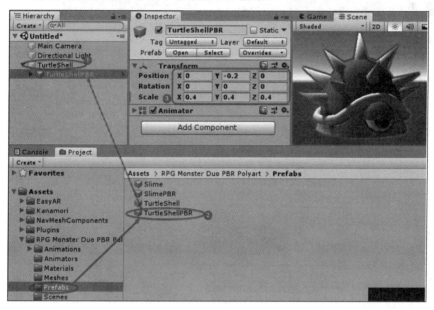

图 8-143

选中 TurtleShellNew 游戏对象,将 Kanamori/Scripts 目录下的 ShowChildObject 拖到其上成为其组件,如图 8-144 所示。

单击 Add Component→Physics→Sphere Collider,为 TurtleShellNew 游戏对象添加一个球体碰撞器,如图 8-145 所示。

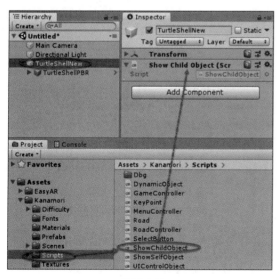

图 8-144

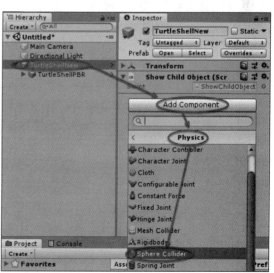

图 8-145

修改球体碰撞器的 Radius 为 0.3，将 TurtleShellNew 游戏对象拖到 Kanamori/Prefabs 目录下成为预制件，如图 8-146 所示。

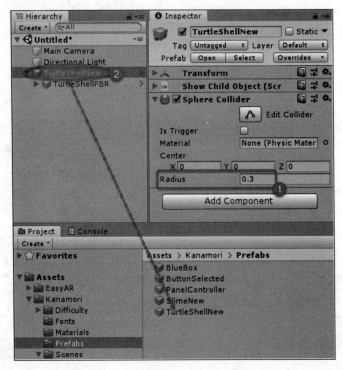

图 8-146

2. 修改选择按钮

修改 SelectButton 脚本，添加一个新的属性 target，用于保存按钮对应的导航目标点。

```
public Transform target;
```

3. 添加加载方法

修改 DbgNavigationController 脚本，添加导航目标点预制件、按钮预制件以及滚动视图内容框属性。

遍历 JSON 列表，根据关键点类型只添加类型为目的地的关键点。在稀疏空间地图下添加导航目标点预制件生成的目标点游戏对象，并将其赋值给添加的按钮的 target 属性。

```
public Transform targetPrefab;
public SelectButton selectButton;
public Transform svContent;
void Start()
{
    ...
    LoadTarget();
    ShowNav();
}
public void LoadTarget()
```

```
{
    var list = game.LoadKeyPoint();
    foreach (var item in list)
    {
        KeyPoint point = JsonUtility.FromJson<KeyPoint>(item);
        if (point.pointType == 0)
        {
            var target = Instantiate(targetPrefab, ssMap);
            target.localPosition = point.position;
            target.GetComponent<ShowChildObject>().SetVisible(false);
            var btn = Instantiate(selectButton, svContent);
            btn.GetComponentInChildren<Text>().text = point.name;
            btn.target = target;
        }
    }
}
```

选中 SceneMaster 游戏对象，将 Content 游戏对象拖到 Sv Content 属性中，将 Kanamori/Prefabs 目录下的 ButtonSelected 游戏对象拖到 Select Button 属性中，将 TurtleShellNew 预制件拖到 Target Prefab 属性中，如图 8-147 所示。

从菜单场景运行，按钮目的地的游戏对象和对应按钮被添加到界面上，如图 8-148 所示。

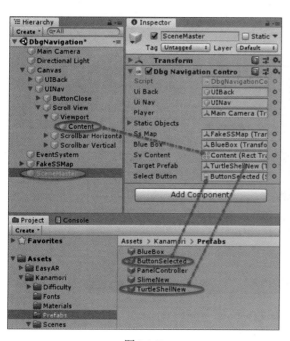

图 8-147

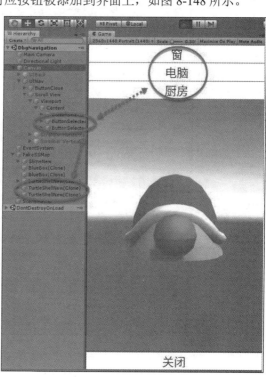

图 8-148

8.6.7 添加路径

1. 添加路径预制件

新建一个场景,单击菜单 GameObject→3D Object→Plane,添加一个平面。修改平面名称为 road,并将其拖到 Kanamori/Prefabs 目录中成为预制件,如图 8-149 所示。

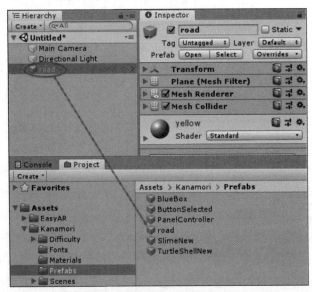

图 8-149

2. 添加路径游戏对象的父节点

在 FakeSSMap 游戏对象下添加一个空的游戏对象并命名为 Roads,如图 8-150 所示。

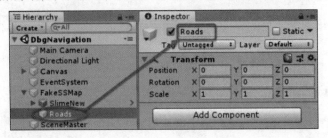

图 8-150

3. 通过修改脚本 DbgNavigationController 添加加载路径方法

```
private void LoadRoad()
{
    var list = game.LoadRoad();
    foreach (var item in list)
    {
        var road = JsonUtility.FromJson<Road>(item);
        var tfRoad = Instantiate(roadPrefab, ssMap.Find("Roads"));
        tfRoad.localPosition = (road.startPosition + road.endPosition) / 2;
```

```
        tfRoad.LookAt(road.startPosition);
        tfRoad.localScale = new Vector3(0.02f, 1f,
        (road.endPosition - road.startPosition).magnitude * 0.1f + 0.2f);
    }
}
```

选中 SceneMaster 游戏对象，将 Kanamori/Prefabs 目录中的 road 预制件拖到 Road Prefab 属性中，如图 8-151 所示。

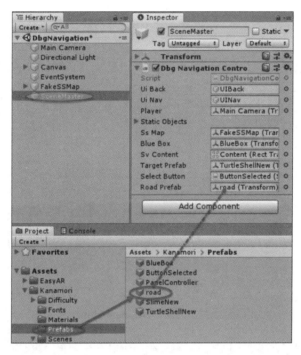

图 8-151

从菜单场景开始运行，能看到在 Roads 游戏对象中添加了路径，如图 8-152 所示。

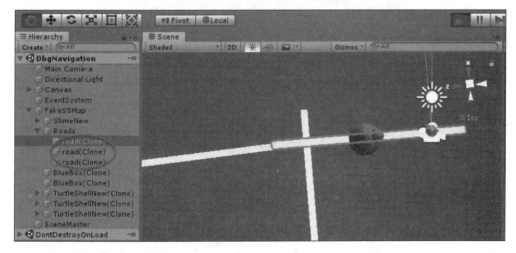

图 8-152

8.6.8 路径导航和显示

1. 添加导航网格表面

选中 Roads 游戏对象，单击 Add Component→Navigation→NavMeshSurface，为游戏对象添加一个导航网格表面组件，如图 8-153 所示。

单击菜单 Window→AI→Navigation，打开导航设置窗口，在窗口中选中 Agents 标签，修改 Radius 为 0.05，如图 8-154 所示。

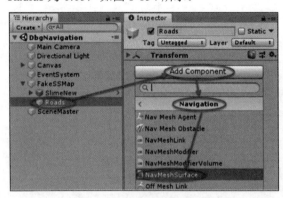 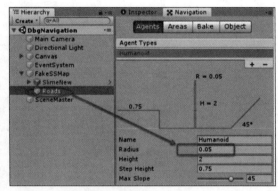

图 8-153　　　　　　　　　　　　　　图 8-154

修改 Collect Objects 属性为 Children，只使用 Roads 游戏对象下的内容来构建；修改 Use Geometry 属性为 Physics Colliders，使用关闭 Renderer 的方法隐藏路径，如图 8-155 所示。

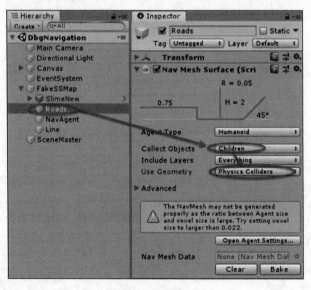

图 8-155

2. 添加导航代理

在 FakeSSMap 游戏对象下，新建一个空的游戏对象并命名为 NavAgent，单击 Add Component →Navigation→Nav Mesh Agent，为游戏对象添加一个导航代理组件，如图 8-156 所示。去掉 Nav Mesh Agent 的激活选项，如图 8-157 所示。

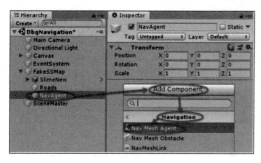
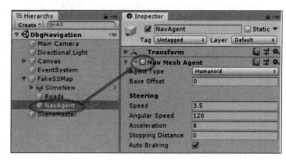

图 8-156   图 8-157

3. 添加路径显示

在 FakeSSMap 游戏对象下新建一个空的游戏对象并命名为 Line，单击 Add Component→Effects→Line Renderer，为游戏对象添加一个线条渲染器组件，如图 8-158 所示。

设置 Materials 的属性只有 1 个 Sprites-Default 元素，修改 Width 属性为 0.000, 0.11，修改 Color 属性为蓝色，如图 8-159 所示。

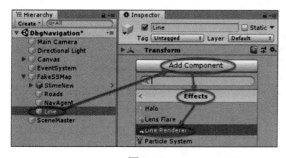
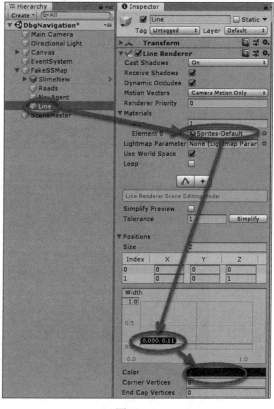

图 8-158   图 8-159

4. 添加动态烘焙脚本

在 DbgNavigationController 脚本中添加导航用的属性，在激活代理后动态烘焙。

```
public NavMeshSurface surface;
public NavMeshAgent agent;
private NavMeshPath path;
void Start()
{
    ...
    BakePath();
    ShowNav();
```

```
}
private void BakePath()
{
    agent.enabled = false;
    surface.BuildNavMesh();
    path = new NavMeshPath();
}
```

选中 SceneMaster 游戏对象，将 Roads 游戏对象拖到 Surface 属性中，将 NavAgent 游戏对象拖到 Agent 属性中，如图 8-160 所示。

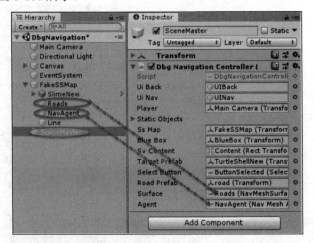

图 8-160

5．添加按钮点击内容

点击按钮以后，重复调用显示路径的方法来显示导航线路。

```
private Transform target;
public LineRenderer lineRenderer;
private void DisplayPath()
{
    agent.transform.position = player.position;
    agent.enabled = true;
    agent.CalculatePath(target.position, path);
    lineRenderer.positionCount = path.corners.Length;
    lineRenderer.SetPositions(path.corners);
    agent.enabled = false;
}
public void SelectButtonClicked(Transform btn)
{
    CancelInvoke("DisplayPath");
    target = btn.GetComponent<SelectButton>().target;
    InvokeRepeating("DisplayPath", 0, refresh);
    Close();
}
```

选中 SceneMaster 游戏对象，将 Line 游戏对象拖到 Line Renderer 属性中，设置 Refresh 属性为 0.1，如图 8-161 所示。

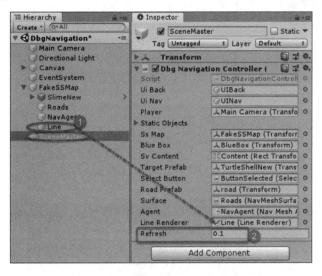

图 8-161

从菜单场景运行，可以看出来导航功能也实现了，如图 8-162 所示。

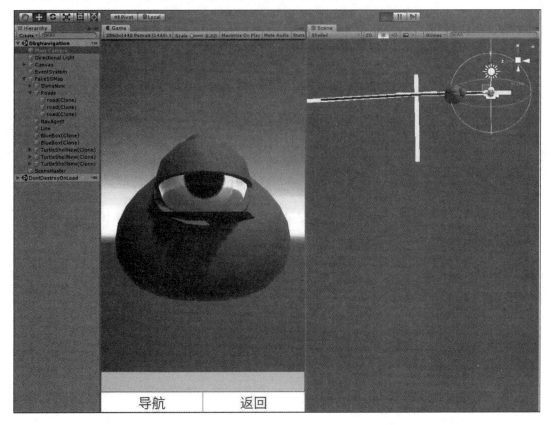

图 8-162

6. 隐藏路径

打开 Kanamori/Prefabs 目录下的 road 预制件，取消 Mesh Renderer 组件的激活，路径就不会被看到了，如图 8-163 所示。

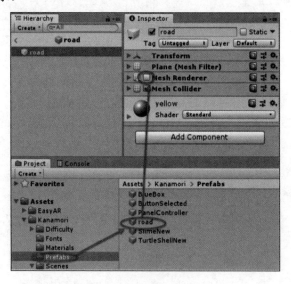

图 8-163

完整的代码如下：

```
using UnityEngine;
using UnityEngine.UI;
using UnityEngine.AI;

namespace Kanamori.Dbg
{
    public class DbgNavigationController : MonoBehaviour
    {
        public GameObject uiBack;
        public GameObject uiNav;
        private GameController game;
        public Transform player;
        public ShowChildObject[] staticObjects;
        public Transform ssMap;
        public Transform blueBox;
        public Transform svContent;
        public Transform targetPrefab;
        public SelectButton selectButton;
        public Transform roadPrefab;
        public NavMeshSurface surface;
        public NavMeshAgent agent;
        private NavMeshPath path;
        private Transform target;
```

```csharp
public LineRenderer lineRenderer;
public float refresh;
void Start()
{
    game = FindObjectOfType<GameController>();
    SetStaticObject();
    LoadObjects();
    LoadTarget();
    LoadRoad();
    BakePath();
    ShowNav();
}
private void SetStaticObject()
{
    foreach (var item in staticObjects)
    {
        item.SetVisible((item.transform.position - player.position)
            .magnitude <= 2);
    }
}
public void Close()
{
    uiBack.SetActive(true);
    uiNav.SetActive(false);
}
public void ShowNav()
{
    uiBack.SetActive(false);
    uiNav.SetActive(true);
}
public void Back()
{
    game.BackDbgMenu();
}
private void LoadObjects()
{
    if (game)
    {
        var list = game.LoadDynamicObject();
        foreach (var item in list)
        {
            var dynamicObject = JsonUtility.FromJson<DynamicObject>(item);
            var tf = Instantiate(blueBox, ssMap);
            tf.localPosition = dynamicObject.position;
            tf.localEulerAngles = dynamicObject.rotation;
            tf.localScale = dynamicObject.scale;
            var obj = tf.GetComponent<ShowSelfObject>();
```

```csharp
                    obj.SetVisible((tf.position - player.position).magnitude <= 2);
                }
            }
        }
        private void LoadTarget()
        {
            if (game)
            {
                var list = game.LoadKeyPoint();
                foreach (var item in list)
                {
                    KeyPoint point = JsonUtility.FromJson<KeyPoint>(item);
                    if (point.pointType == 0)
                    {
                        var target = Instantiate(targetPrefab, ssMap);
                        target.localPosition = point.position;
                        target.GetComponent<ShowChildObject>().SetVisible(false);
                        var btn = Instantiate(selectButton, svContent);
                        btn.GetComponentInChildren<Text>().text = point.name;
                        btn.target = target;
                    }
                }
            }
        }
        private void LoadRoad()
        {
            if (game)
            {
                var list = game.LoadRoad();
                foreach (var item in list)
                {
                    var road = JsonUtility.FromJson<Road>(item);
                    var tfRoad = Instantiate(roadPrefab, ssMap.Find("Roads"));
                    tfRoad.localPosition = (road.startPosition + road.endPosition) / 2;
                    tfRoad.LookAt(road.startPosition);
                    tfRoad.localScale = new Vector3(0.02f, 1f, (road.endPosition - road.startPosition).magnitude * 0.1f + 0.2f);
                }
            }
        }
        private void BakePath()
        {
            agent.enabled = false;
            surface.BuildNavMesh();
            path = new NavMeshPath();
        }
        private void DisplayPath()
```

```
        {
            agent.transform.position = player.position;
            agent.enabled = true;
            agent.CalculatePath(target.position, path);
            lineRenderer.positionCount = path.corners.Length;
            lineRenderer.SetPositions(path.corners);
            agent.enabled = false;
        }
        public void SelectButtonClicked(Transform btn)
        {
            CancelInvoke("DisplayPath");
            target = btn.GetComponent<SelectButton>().target;
            InvokeRepeating("DisplayPath", 0, refresh);
            Close();
        }
    }
}
```

第 9 章
◀ 实际场景开发 ▶

方便用来调试的场景开发完成以后，现在可以开始实际场景的开发了。

打开之前开发的测试场景，将其另存到 Kanamori/Scenes 目录下并去掉 Dbg 前缀。保存以后，单击菜单 File→Build Settings，打开 Build Settings 窗口，删除 Scenes In Build 下的原有场景，将新的场景拖到 Scenes In Build 下，确保 Menu 场景是第一个场景，如图 9-1 所示。

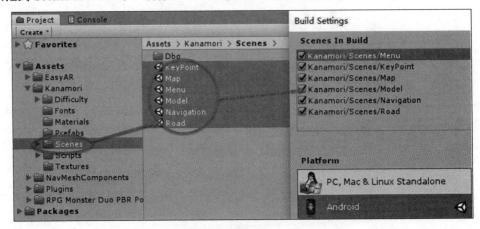

图 9-1

9.1 菜单场景开发

菜单场景修改比较简单，只需要修改原有的参数就可以了，如图 9-2 所示。按钮参数如表 9-1 所示。

表 9-1 按钮参数

按 钮	参 数
ButtonNavi	Navigation
ButtonPoint	KeyPoint
ButtonRoad	Road
ButtonModel	Model
ButtonAddMap	Map

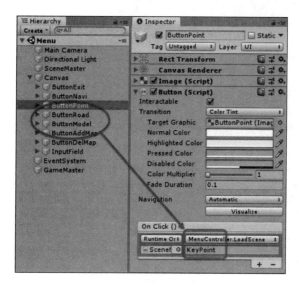

图 9-2

将目录 Plugins/IngameDebugConsole 下的 IngameDebugConsole 预制件拖到场景中，如图 9-3 所示。这时场景中会多出一个小贴纸，点击它后能看到控制台的输出信息，这在调试的时候很方便。

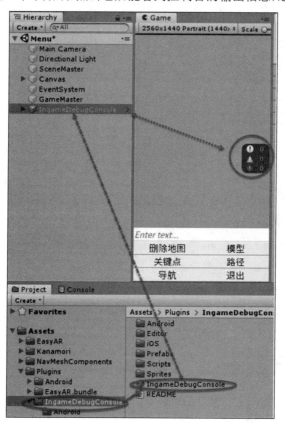

图 9-3

9.2 地图场景开发

9.2.1 添加稀疏空间地图游戏对象

选中 Main Camera 游戏对象，修改 Clear Flags 属性为 Solid Color，如图 9-4 所示。

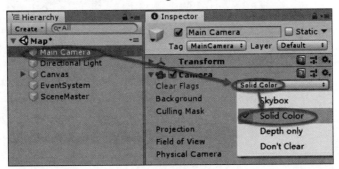

图 9-4

将 EasyAR/Prefabs/Primitives 目录下的 WorldRoot 预制件和 SparseSpatialMap 预制件拖到场景中，如图 9-5 所示。

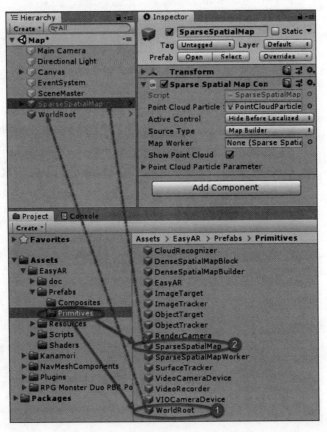

图 9-5

将 EasyAR/Prefabs/Composites 目录下的 EasyAR_SparseSpatialMapWorker 预制件拖到场景中，将 WorldRoot 游戏对象拖到 EasyAR_SparseSpatialMapWorker 游戏对象的 World Root Controller 属性中，如图 9-6 所示。

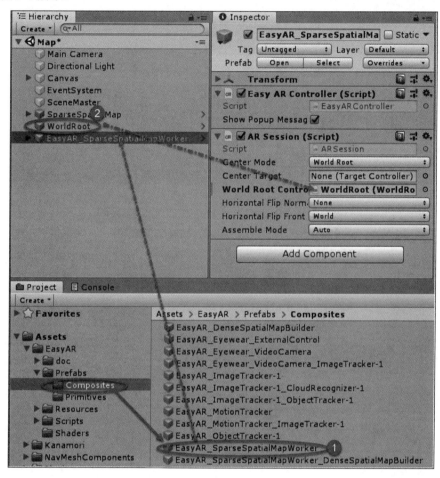

图 9-6

9.2.2 修改返回功能

1. 修改游戏控制脚本

在 GameController 脚本中，添加返回到实际菜单场景的方法。

```
public void BackMenu(){
    SceneManager.LoadScene("Menu");
}
```

2. 修改地图场景脚本

打开 DbgMapController 脚本，将其另存到 Kanamori/Scripts 目录下，名称为 MapController。修改脚本的命名空间为 Kanamori，修改类名为 MapController。

```csharp
namespace Kanamori
{
    public class MapController : MonoBehaviour
    ...
}
```

修改返回方法:

```csharp
public void Back()
{
    if (game)
    {
        game.BackMenu();
    }
}
```

此时，GameController 脚本已经修改完成，完整的内容如下:

```csharp
using System.Collections.Generic;
using UnityEngine;
using UnityEngine.SceneManagement;
using System.IO;
using System;

namespace Kanamori
{
    public class GameController : MonoBehaviour
    {
        private static GameController instance = null;
        public string inputName;
        private static readonly string pathDynamicObject = "/dynamicobject.txt";
        private static readonly string pathKeyPoint = "/keypoint.txt";
        private static readonly string pathRoad = "/road.txt";
        void Awake()
        {
            //实现单实例
            if (instance == null)
            {
                instance = this;
                DontDestroyOnLoad(gameObject);
            }
            else if (this != instance)
            {
                Destroy(gameObject);
                return;
            }
        }
        public string GetMapName()
```

```csharp
        {
            return PlayerPrefs.GetString("MapName", "");
        }
        public void SaveMapName(string mapName)
        {
            PlayerPrefs.SetString("MapName", mapName);
        }
        public string GetMapID()
        {
            return PlayerPrefs.GetString("MapID", "");
        }
        public void SaveMapID(string mapID)
        {
            PlayerPrefs.SetString("MapID", mapID);
        }
        public void DelMap()
        {
            PlayerPrefs.DeleteKey("MapID");
            PlayerPrefs.DeleteKey("MapName");
        }
        public void BackDbgMenu()
        {
            SceneManager.LoadScene("DbgMenu");
        }
        public void BackMenu(){
            SceneManager.LoadScene("Menu");
        }
        public void SaveRoad(string[] stringArray)
        {
            SaveStringArray(stringArray, Application.persistentDataPath + pathRoad);
        }
        public List<string> LoadRoad()
        {
            return LoadStringList(Application.persistentDataPath + pathRoad);
        }
        public void SaveKeyPoint(string[] stringArray)
        {
            SaveStringArray(stringArray, Application.persistentDataPath + pathKeyPoint);
        }
        public List<string> LoadKeyPoint()
        {
            return LoadStringList(Application.persistentDataPath + pathKeyPoint);
        }
        public void SaveDynamicObject(string[] stringArray)
        {
```

```csharp
            SaveStringArray(stringArray, Application.persistentDataPath +
pathDynamicObject);
        }
        public List<string> LoadDynamicObject()
        {
            return LoadStringList(Application.persistentDataPath +
pathDynamicObject);
        }
        private void SaveStringArray(string[] stringArray, string path)
        {
            try
            {
                using (StreamWriter writer = new StreamWriter(path))
                {
                    foreach (var s in stringArray)
                    {
                        writer.WriteLine(s);
                    }
                }
            }
            catch (Exception ex)
            {
                Debug.Log(ex.Message);
            }
        }
        private List<string> LoadStringList(string path)
        {
            List<string> list = new List<string>();
            try
            {
                using (StreamReader reader = new StreamReader(path))
                {
                    while (!reader.EndOfStream)
                    {
                        list.Add(reader.ReadLine());
                    }
                }
            }
            catch (Exception ex)
            {
                Debug.Log(ex.Message);
            }
            return list;
        }
    }
}
```

9.2.3 添加保存功能

1. 添加保存按钮控制

保存按钮只有在稀疏空间地图正确跟踪以后才能被点击。添加稀疏空间地图和保存按钮的公开属性：

```
public ARSession session;
public SparseSpatialMapWorkerFrameFilter mapWorker;
public SparseSpatialMapController map;
public Button btnSave;
```

在 Start 方法中，监听跟踪状态并控制保存按钮是否能点击：

```
void Start()
{
    game = FindObjectOfType<GameController>();
    btnSave.interactable = false;
    session.WorldRootController.TrackingStatusChanged += OnTrackingStatusChanged;
    if (session.WorldRootController.TrackingStatus == MotionTrackingStatus.Tracking)
    {
        btnSave.interactable = true;
    }
    else
    {
        btnSave.interactable = false;
    }
}
private void OnTrackingStatusChanged(MotionTrackingStatus status)
{
    if (status == MotionTrackingStatus.Tracking)
    {
        btnSave.interactable = true;
        text.text="进入跟踪状态。";
    }
    else
    {
        btnSave.interactable = false;
        text.text="跟踪状态异常";
    }
}
```

2. 修改保存地图的方法

```
public void SaveMap()
{
    btnSave.interactable = false;
```

```
//地图保存结果反馈
mapWorker.BuilderMapController.MapHost += (mapInfo, isSuccess, error) =>
{
    if (isSuccess)
    {
        game.SaveMapID(mapInfo.ID);
        game.SaveMapName(mapInfo.Name);
        text.text = "地图保存成功。";
    }
    else
    {
        text.text = "地图保存出错：" + error;
        btnSave.interactable = true;
    }
};
try
{
    //保存地图
    mapWorker.BuilderMapController.Host(game.inputName, null);
    text.text = "开始保存地图，请稍等。";
}
catch (Exception ex)
{
    text.text = "保存出错：" + ex.Message;
    btnSave.interactable = true;
}
```

3. 设置界面

选中 SceneMaster 游戏对象，删除原有的 Dbg Map Controller 脚本组件。将 Kanamori/Scripts 目录下的 MapController 脚本拖到 SceneMaster 游戏对象中并成为其组件，如图 9-7 所示。其中，游戏对象对应的属性如表 9-2 所示。

表 9-2　为脚本属性赋值

游戏对象	属　　性
ButtonSave	Btn Save
Text	Text
EasyAR_SparseSpatialMapWorker	Session
SparseSpatialMapWorker	Map Worker
SparseSpatialMap	Map

设置 ButtonBack 按钮的点击事件为 SceneMaster 游戏对象上 MapController 脚本下的 Back 方法。设置 ButtonSave 按钮的点击事件为 SceneMaster 游戏对象上 MapController 脚本下的 SaveMap 方法，如图 9-8 所示。

第 9 章 实际场景开发

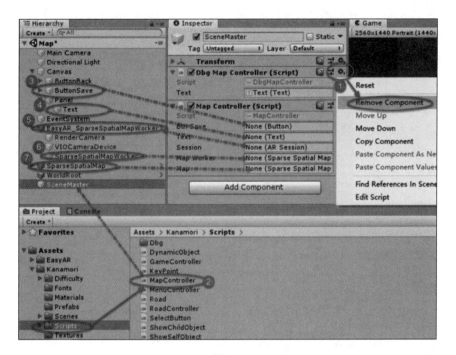

图 9-7

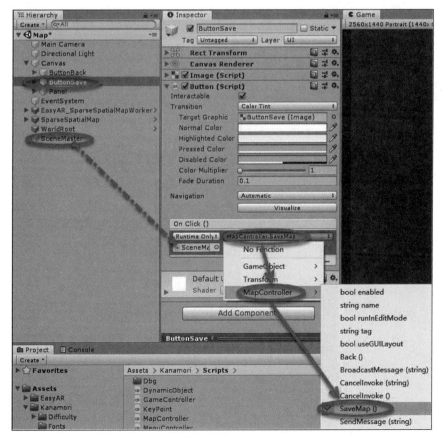

图 9-8

273

完整的代码如下：

```csharp
using UnityEngine;
using UnityEngine.UI;
using easyar;
using System;

namespace Kanamori
{
    public class MapController : MonoBehaviour
    {
        private GameController game;
        public Button btnSave;
        public Text text;
        public ARSession session;
        public SparseSpatialMapWorkerFrameFilter mapWorker;
        public SparseSpatialMapController map;
        void Start()
        {
            game = FindObjectOfType<GameController>();
            btnSave.interactable = false;
            session.WorldRootController.TrackingStatusChanged += OnTrackingStatusChanged;
            if (session.WorldRootController.TrackingStatus == MotionTrackingStatus.Tracking)
            {
                btnSave.interactable = true;
            }
            else
            {
                btnSave.interactable = false;
            }
        }
        private void OnTrackingStatusChanged(MotionTrackingStatus status)
        {
            if (status == MotionTrackingStatus.Tracking)
            {
                btnSave.interactable = true;
                text.text = "进入跟踪状态。";
            }
            else
            {
                btnSave.interactable = false;
                text.text = "跟踪状态异常";
            }
        }
```

```
        public void SaveMap()
        {
            btnSave.interactable = false;
            //地图保存结果反馈
            mapWorker.BuilderMapController.MapHost += (mapInfo, isSuccess, error) =>
            {
                if (isSuccess)
                {
                    game.SaveMapID(mapInfo.ID);
                    game.SaveMapName(mapInfo.Name);
                    text.text = "地图保存成功。";
                }
                else
                {
                    text.text = "地图保存出错: " + error;
                    btnSave.interactable = true;
                }
            };
            try
            {
                //保存地图
                mapWorker.BuilderMapController.Host(game.inputName, null);
                text.text = "开始保存地图, 请稍等。";
            }
            catch (Exception ex)
            {
                text.text = "保存出错: " + ex.Message;
                btnSave.interactable = true;
            }
        }
        /// <summary>
        /// 返回
        /// </summary>
        public void Back()
        {
            if (game)
            {
                game.BackMenu();
            }
        }
    }
}
```

9.3 模型场景开发

9.3.1 场景设置

1. 添加稀疏空间地图游戏对象

选中 Main Camera 游戏对象，修改 Clear Flags 属性为 Solid Color。将 EasyAR/Prefabs/Primitives 目录下的 WorldRoot 预制件和 SparseSpatialMap 预制件拖到场景中。将 EasyAR/Prefabs/Composites 目录下的 EasyAR_SparseSpatialMapWorker 预制件拖到场景中。将 WorldRoot 游戏对象拖到 EasyAR_SparseSpatialMapWorker 游戏对象的 World Root Controller 属性中，如图 9-9 所示。

2. 设置地图

选中 SparseSpatialMap 游戏对象，修改 Source Type 属性为 Map Manager，删除 FakeSSMap 游戏对象，如图 9-10 所示。

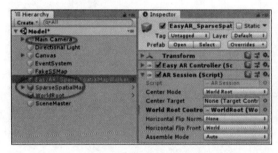
图 9-9

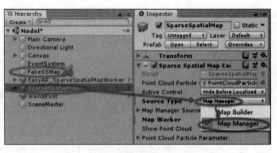
图 9-10

选中 SparseSpatialMapWorker 游戏对象，设置 Localization Mode 为 Keep Update，如图 9-11 所示。

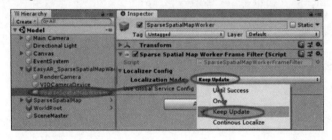
图 9-11

9.3.2 脚本修改

1. 另存脚本

打开 DbgModelController 脚本，将其另存到 Kanamori/Scripts 目录下，名称为 ModelController。修改脚本的命名空间为 Kanamori，修改类名为 ModelController：

```
namespace Kanamori
{
```

```
public class ModelController : MonoBehaviour
    ...
}
```

2. 修改返回方法

修改返回的方法：

```
public void Back()
{
    if (game)
    {
        game.BackMenu();
    }
}
```

3. 添加稀疏空间地图公开属性

添加 3 个相关的属性以及按钮的属性：

```
public Button btnAdd;
public ARSession session;
public SparseSpatialMapWorkerFrameFilter mapWorker;
public SparseSpatialMapController map;

void Start()
{
    game = FindObjectOfType<GameController>();
    uiControl = FindObjectOfType<UIControlObject>();
    btnAdd.interactable = false;
    ...
}
```

4. 添加加载地图方法

加载地图，返回信息并控制"添加"按钮是否能点击：

```
void Start()
{
    ...
    LoadMap();
}
private void LoadMap()
{
    //设置地图
    map.MapManagerSource.ID = game.GetMapID();
    map.MapManagerSource.Name = game.GetMapName();
    //地图获取反馈
    map.MapLoad += (map, status, error) =>
    {
```

```
            if (status)
            {
                textShow.text = "地图加载成功。";
            }
            else
            {
                textShow.text = "地图加载失败：" + error;
            }
        };
        //定位成功事件
        map.MapLocalized += () =>
        {
            textShow.text = "稀疏空间定位成功。";
            btnAdd.interactable = true;
        };
        //停止定位事件
        map.MapStopLocalize += () =>
        {
            textShow.text = "停止稀疏空间定位。";
            btnAdd.interactable = false;
        };
        textShow.text = "开始本地化稀疏空间。";
        mapWorker.Localizer.startLocalization();       //本地化地图
}
```

5. 脚本设置

选中 SceneMaster 游戏对象，删除原有的 Dbg Model Controller 脚本组件，将 Kanamori/Scripts 目录下的 ModelController 脚本拖到 SceneMaster 游戏对象上成为其组件，设置 SceneMaster 游戏对象的属性（见表 9-3），将 Kanamori/Prefabs 目录下的 BlueBox 预制件拖到 Blue Box 属性中为其赋值，如图 9-12 所示。

表 9-3　为脚本属性赋值

游戏对象	属　　性
Sphere	Front Camera
AddUI	Add UI
ButtonAdd	Btn Add
SaveUI	Save UI
Text	Text Show
EasyAR_SparseSpatialMapWorker	Session
SparseSpatialMapWorker	Map Worker
SparseSpatialMap	Map

设置按钮点击事件对应的方法（都来自 SceneMaster 游戏对象上的 ModelController 脚本，见表 9-4），如图 9-13 所示。

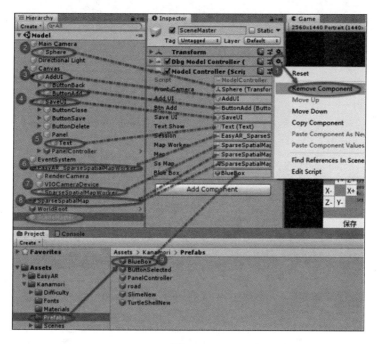

图 9-12

表 9-4 按钮对应方法

游戏对象	方　　法
ButtonBack	Back
ButtonAdd	Add
ButtonClose	Close
ButtonSave	Save
ButtonDelete	Delete

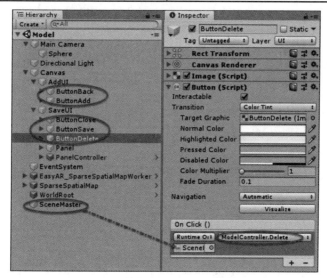

图 9-13

完整的代码如下：

```csharp
using UnityEngine;
using UnityEngine.UI;
using UnityEngine.EventSystems;
using easyar;

namespace Kanamori
{
    public class ModelController : MonoBehaviour
    {
        private GameController game;
        public Transform frontCamera;
        public GameObject addUI;
        public Button btnAdd;
        public GameObject saveUI;
        private UIControlObject uiControl;
        public Text textShow;
        public SparseSpatialMapWorkerFrameFilter mapWorker;
        public SparseSpatialMapController map;
        public Transform ssMap;
        public GameObject blueBox;
        void Start()
        {
            game = FindObjectOfType<GameController>();
            uiControl = FindObjectOfType<UIControlObject>();
            btnAdd.interactable = false;
            Close();
            Load();
            LoadMap();
        }
        private void LoadMap()
        {
            //设置地图
            map.MapManagerSource.ID = game.GetMapID();
            map.MapManagerSource.Name = game.GetMapName();
            //地图获取反馈
            map.MapLoad += (map, status, error) =>
            {
                if (status)
                {
                    textShow.text = "地图加载成功。";
                }
                else
                {
                    textShow.text = "地图加载失败：" + error;
                }
```

```csharp
        };
        //定位成功事件
        map.MapLocalized += () =>
        {
            textShow.text = "稀疏空间定位成功。";
            btnAdd.interactable = true;
        };
        //停止定位事件
        map.MapStopLocalize += () =>
        {
            textShow.text = "停止稀疏空间定位。";
            btnAdd.interactable = false;
        };
        textShow.text = "开始本地化稀疏空间。";
        mapWorker.Localizer.startLocalization();       //本地化地图
    }
    public void Back()
    {
        if (game)
        {
            game.BackMenu();
        }
    }
    public void Add()
    {
        var tf = Instantiate(blueBox, ssMap).transform;
        tf.position = frontCamera.position;
    }
    public void Close()
    {
        addUI.SetActive(true);
        saveUI.SetActive(false);
    }
    void Update()
    {
        if (Application.platform == RuntimePlatform.WindowsEditor)
        {
            if (Input.GetMouseButtonDown(0)
                && !EventSystem.current.IsPointerOverGameObject())
            {
                Ray ray = Camera.main.ScreenPointToRay(Input.mousePosition);
                TouchedObject(ray);
            }
        }
        else
        {
            if (Input.touchCount == 1
```

```csharp
                    && Input.touches[0].phase == TouchPhase.Began
                    && !EventSystem.current.IsPointerOverGameObject (Input.touches[0].fingerId))
            {
                Ray ray = Camera.main.ScreenPointToRay (Input.touches[0].position);
                TouchedObject(ray);
            }
        }
    }
    private void TouchedObject(Ray ray)
    {
        if (Physics.Raycast(ray, out RaycastHit hit))
        {
            addUI.SetActive(false);
            saveUI.SetActive(true);
            uiControl.SetSelected(hit.transform);
            textShow.text = "选中物体";
        }
    }
    public void Save()
    {
        string[] jsons = new string[ssMap.childCount - 1];
        for (int i = 0; i < ssMap.childCount; i++)
        {
            if (ssMap.GetChild(i).name != "PointCloudParticleSystem")
            {
                DynamicObject dynamicObject = new DynamicObject();
                dynamicObject.position = ssMap.GetChild(i).localPosition;
                dynamicObject.rotation = ssMap.GetChild(i).localEulerAngles;
                dynamicObject.scale = ssMap.GetChild(i).localScale;
                jsons[i - 1] = JsonUtility.ToJson(dynamicObject);
            }
        }
        if (game)
        {
            game.SaveDynamicObject(jsons);
            textShow.text = "保存" + (ssMap.childCount - 1) + "个游戏对象";
        }
    }
    public void Delete()
    {
        var go = uiControl.selected.gameObject;
        uiControl.ClearSelected();
        Destroy(go);
        textShow.text = "删除选中物体，请保存结果。";
    }
```

```
    private void Load()
{
   if (game)
   {
      var list = game.LoadDynamicObject();
      foreach (var item in list)
      {
         var dynamicObject = JsonUtility.FromJson<DynamicObject>(item);
         var tf = Instantiate(blueBox, ssMap).transform;
         tf.localPosition = dynamicObject.position;
         tf.localEulerAngles = dynamicObject.rotation;
         tf.localScale = dynamicObject.scale;
      }
   }
}
```

9.4 关键点场景开发

9.4.1 添加平面跟踪图像

在场景中新建 StreamingAssets 目录，并将识别图像拖到该目录下，同时将识别图像打印出来，如图 9-14 所示。

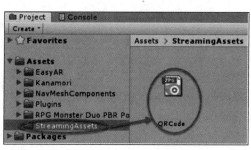

图 9-14

选中 Main Camera 游戏对象，修改 Clear Flags 属性为 Solid Color。将 EasyAR/Prefabs/Primitives 目录下的 WorldRoot 预制件、SparseSpatialMap 预制件和 ImageTarget 预制件拖到场景中。将 EasyAR/Prefabs/Composites 目录下的 EasyAR_SparseSpatialMapWorker_ImageTracker-1 预制件拖到场景中。将 WorldRoot 游戏对象拖到 EasyAR_SparseSpatialMapWorker 游戏对象的 World Root Controller 属性中并为其赋值。

删除原有的 FakeSSMap 游戏对象和 FakeImageTarget 游戏对象；根据图像的大小设置 ImageTarget 属性，这里设置 Path 为 QRCode.jpg、Scale 为 0.173，如图 9-15 所示。

在 ImageTarget 游戏对象下添加一个方块，方块位置在"0，0，0"，如图 9-16 所示。

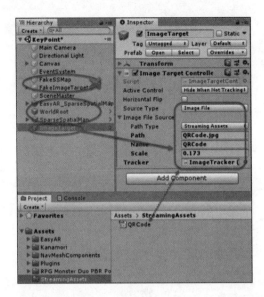

图 9-15

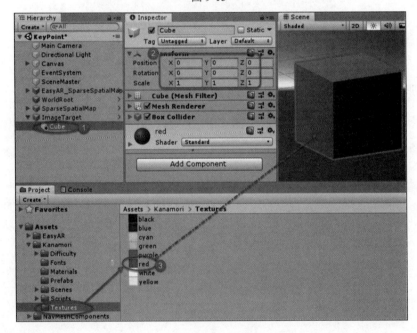

图 9-16

9.4.2 脚本准备

打开 DbgKeyPointController 脚本,将其另存到 Kanamori/Scripts 目录下,名称为 KeyPointController。修改脚本的命名空间为 Kanamori,修改类名为 KeyPointController。

删除脚本原有的 cam 属性和 Update 方法中的 LookAt 语句。这两个内容是为了调试用的,不能出现在实际使用场景中,否则会出错。

```
public Transform cam;
void Update()
```

```
{
    cam.LookAt(image);
    ...
}
```

选中 SceneMaster 游戏对象，删除原有的 Dbg KeyPoint Controller 脚本组件。将 Kanamori/Prefabs 目录下的 ButtonSelected 预制件拖到 Prefab 属性中，将 ImageTarget 游戏对象拖到 Image 属性中，其他属性设置（见表 9-5）按照调试用关键点场景的脚本进行，如图 9-17 所示。

表 9-5　游戏对象对应属性

游戏对象	属　　性
ButtonBack	Ui Back
Panel	Ui Main
ButtonAdd	Btn Add
Text	Text Info
InputField	Input Field
Dropdown	Dropdown
ButtonDelete	Btn Delete

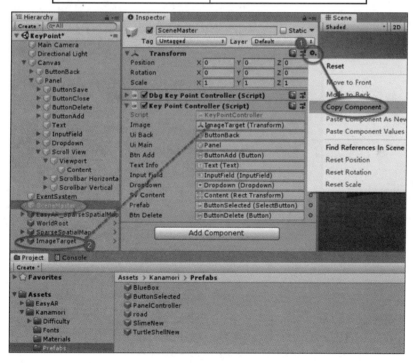

图 9-17

9.4.3　脚本修改

这里的脚本和模型场景的脚本很接近。区别只是在模型脚本中，当定位停止的时候，不能单击按钮，这里是定位停止的时候不能进行平面图像跟踪。

添加对应的属性和方法，用 ImageTrackerFrameFilter 类的 enabled 属性来控制是否进行平面跟踪。原有的 Transform 类型的 map 属性用稀疏空间地图的属性替换。

```csharp
public ARSession session;
public SparseSpatialMapWorkerFrameFilter mapWorker;
public SparseSpatialMapController map;
public ImageTrackerFrameFilter imageTracker;
void Start()
{
    ...
    imageTracker.enabled = false;
    LoadMap();
}
private void LoadMap()
{
    //设置地图
    map.MapManagerSource.ID = game.GetMapID();
    map.MapManagerSource.Name = game.GetMapName();
    //地图获取反馈
    map.MapLoad += (map, status, error) =>
    {
        if (status)
        {
            textInfo.text = "地图加载成功。";
        }
        else
        {
            textInfo.text = "地图加载失败：" + error;
        }
    };
    //定位成功事件
    map.MapLocalized += () =>
    {
        textInfo.text = "稀疏空间定位成功。";
        imageTracker.enabled = true;
    };
    //停止定位事件
    map.MapStopLocalize += () =>
    {
        textInfo.text = "停止稀疏空间定位。";
        imageTracker.enabled = false;
    };
    textInfo.text = "开始本地化稀疏空间。";
    mapWorker.Localizer.startLocalization();      //本地化地图
}
```

游戏对象对应脚本属性如表 9-6 所示。

表 9-6 设置脚本属性

游戏对象	属 性
ImageTracker	Image Tracker
EasyAR_SparseSpatialMapWorker	Session
SparseSpatialMapWorker	Map Worker
SparseSpatialMap	Map

选中 SceneMaster 游戏对象，设置脚本属性，如图 9-18 所示。

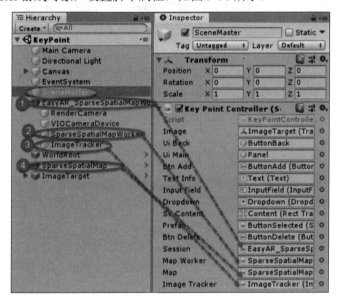

图 9-18

最后，修改返回方法：

```
public void Back()
{
    game.BackMenu();
}
```

设置按钮点击事件对应的方法（都是来自 SceneMaster 游戏对象上的 KeyPointController 脚本的方法，见表 9-7），如图 9-19 所示。

表 9-7 按钮对应的方法

游戏对象	方 法
ButtonBack	Back
ButtonAdd	Add
ButtonClose	Close
ButtonSave	Save
ButtonDelete	Delete

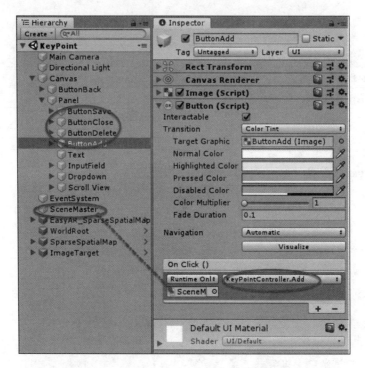

图 9-19

完整的代码如下:

```
using UnityEngine;
using UnityEngine.EventSystems;
using UnityEngine.UI;
using easyar;

namespace Kanamori
{
    public class KeyPointController : MonoBehaviour
    {
        private GameController game;
        public GameObject uiBack;
        public GameObject uiMain;
        private Transform selected;
        public Button btnAdd;
        public Text textInfo;
        public InputField inputField;
        public Dropdown dropdown;
        public Transform svContent;
        public SelectButton prefab;
        public Button btnDelete;
        public SparseSpatialMapWorkerFrameFilter mapWorker;
        public SparseSpatialMapController map;
        public ImageTrackerFrameFilter imageTracker;
```

```csharp
void Start()
{
    game = FindObjectOfType<GameController>();
    Load();
    btnAdd.interactable = false;
    btnDelete.interactable = false;
    Close();
    imageTracker.enabled = false;
    LoadMap();
}
void Update()
{
    if (Application.platform == RuntimePlatform.WindowsEditor)
    {
        if (Input.GetMouseButtonDown(0)
        && !EventSystem.current.IsPointerOverGameObject())
        {
            Ray ray = Camera.main.ScreenPointToRay(Input.mousePosition);
            TouchedObject(ray);
        }
    }
    else
    {
        if (Input.touchCount == 1
        && Input.touches[0].phase == TouchPhase.Began
        && !EventSystem.current.IsPointerOverGameObject(Input.touches[0].fingerId))
        {
            Ray ray = Camera.main.ScreenPointToRay(Input.touches[0].position);
            TouchedObject(ray);
        }
    }
}
private void LoadMap()
{
    //设置地图
    map.MapManagerSource.ID = game.GetMapID();
    map.MapManagerSource.Name = game.GetMapName();
    //地图获取反馈
    map.MapLoad += (map, status, error) =>
    {
        if (status)
        {
            textInfo.text = "地图加载成功。";
        }
        else
```

```csharp
            {
                textInfo.text = "地图加载失败:" + error;
            }
        };
        //定位成功事件
        map.MapLocalized += () =>
        {
            textInfo.text = "稀疏空间定位成功。";
            imageTracker.enabled = true;
        };
        //停止定位事件
        map.MapStopLocalize += () =>
        {
            textInfo.text = "停止稀疏空间定位。";
            imageTracker.enabled = false;
        };
        textInfo.text = "开始本地化稀疏空间。";
        mapWorker.Localizer.startLocalization();    //本地化地图
    }
    private void TouchedObject(Ray ray)
    {
        if (Physics.Raycast(ray, out RaycastHit hit))
        {
            uiBack.SetActive(false);
            uiMain.SetActive(true);
            var tf = new GameObject().transform;
            tf.position = hit.transform.position;
            tf.parent = map.transform;
            selected = tf;
            btnAdd.interactable = true;
        }
    }
    public void Back()
    {
        if (game)
        {
            game.BackMenu();
        }
    }
    public void Close()
    {
        uiMain.SetActive(false);
        uiBack.SetActive(true);
    }
    public void Add()
    {
        if (!string.IsNullOrEmpty(inputField.text) && selected != null)
```

```csharp
        {
            SelectButton btn = Instantiate(prefab, svContent);

            btn.keyPoint.name = inputField.text;
            btn.keyPoint.position = selected.localPosition;
            btn.keyPoint.pointType = dropdown.value;

            btn.GetComponentInChildren<Text>().text = inputField.text;

            inputField.text = "";
            selected = null;
            textInfo.text = "添加完成。";
            btnAdd.interactable = false;
        }
    }
    public void Save()
    {
        string[] jsons = new string[svContent.childCount];
        for (int i = 0; i < svContent.childCount; i++)
        {
            jsons[i] =
JsonUtility.ToJson(svContent.GetChild(i).GetComponent<SelectButton>().keyPoint);
        }
        if (game)
        {
            game.SaveKeyPoint(jsons);
            textInfo.text = "保存完成。";
        }
    }
    private void Load()
    {
        if (game)
        {
            var list = game.LoadKeyPoint();
            foreach (var item in list)
            {
                SelectButton btn = Instantiate(prefab, svContent);
                btn.keyPoint = JsonUtility.FromJson<KeyPoint>(item);
                btn.GetComponentInChildren<Text>().text = btn.keyPoint.name;
            }
        }
    }
    public void SelectButtonClicked(Transform btn)
    {
        selected = btn;
        textInfo.text = btn.GetComponentInChildren<Text>().text;
        btnDelete.interactable = true;
```

```
            btnAdd.interactable = false;
    }
    public void Delete()
    {
        Destroy(selected.gameObject);
        textInfo.text = "删除完成。";
        btnDelete.interactable = false;
    }
}
```

9.5 路径场景开发

路径场景是最容易修改的，只需替换一下方法即可。
在 RoadController 脚本中添加返回菜单的方法：

```
public void BackMenu(){
    game.BackMenu();
}
```

选中 ButtonBack 游戏对象，将原有的响应方法改为新添加的方法 BackMenu，如图 9-20 所示。

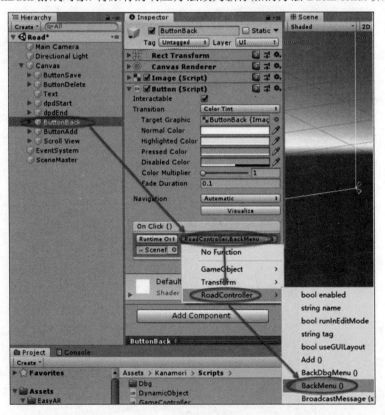

图 9-20

9.6 导航场景开发

9.6.1 设置场景

1. 添加稀疏空间地图内容

这一步和模型场景的实现方法一样。

选中 Main Camera 游戏对象，修改 Clear Flags 属性为 Solid Color。将 EasyAR/Prefabs/Primitives 目录下的 WorldRoot 和 SparseSpatialMap 预制件拖到场景中，将 EasyAR/Prefabs/Composites 目录下的 EasyAR_SparseSpatialMapWorker 预制件拖到场景中，将 WorldRoot 游戏对象拖到 EasyAR_SparseSpatialMapWorker 游戏对象的 World Root Controller 属性中，选中 SparseSpatialMap 游戏对象，修改 Source Type 属性为 Map Manager，如图 9-21 所示。选中 SparseSpatialMapWorker 游戏对象，设置 Localization Mode 为 Keep Update。

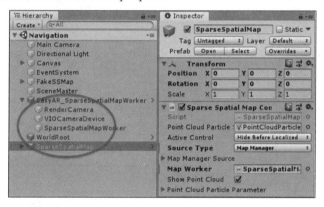

图 9-21

2. 修改稀疏空间地图

将 FakeSSMap 游戏对象下的内容拖到 SparseSpatialMap 下，删除 FakeSSMap 游戏对象；选中 SparseSpatialMap 游戏对象，取消 Show Point Cloud 选项，这里不需要显示特征点，如图 9-22 所示。

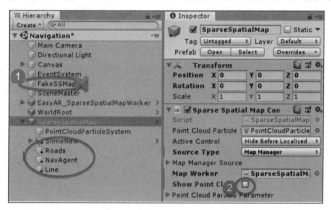

图 9-22

3. 修改脚本及设置

打开 DbgNavigationController 脚本，另存到 Kanamori/Scripts 目录下，名称为 NavigationController。修改脚本的命名空间为 Kanamori，修改类名为 NavigationController。

```
namespace Kanamori
{
    public class NavigationController : MonoBehaviour
    ...
}
```

选中 SceneMaster 游戏对象，删除原有的 Dbg Navigation Controller 脚本组件。将 Kanamori/Scripts 目录下的 NavigationController 脚本拖到 SceneMaster 游戏对象上。设置场景中游戏对象对应脚本属性，如表 9-8 所示。

表 9-8 根据原有脚本为新脚本属性赋值

游戏对象	属　性
UIBack	Ui Back
UINav	Ui Nav
Content	Sv Content
Roads	Surface
NavAgent	Agent
Line	Line Renderer
Main Camera	Player

设置脚本属性对应项目中的预制件，如表 9-9 所示。

表 9-9 设置预制件对应属性

Prefab 预制件	属　性
SlimeNew	Static Objects
BlueBox	Blue Box
TurtleShellNew	Traget Prefab
ButtonSelected	Select Button
road	Road Prefab

将 SparseSpatialMap 游戏对象拖到 Ss Map 属性中为其赋值；Refresh 属性可以略大于原有值，这里设置为 0.5，如图 9-23 所示。

修改返回方法：

```
public void Back()
{
    game.BackMenu();
}
```

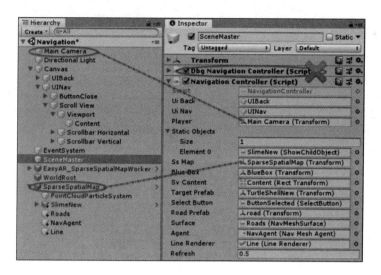

图 9-23

4．设置按钮事件

设置按钮点击事件对应的方法（来自 SceneMaster 游戏对象上的 NavigationController 脚本，见表 9-10），如图 9-24 所示。

表 9-10 设置按钮对应方法

游戏对象	方　　法
ButtonBack	Back
ButtonNav	ShowNav
ButtonClose	Close

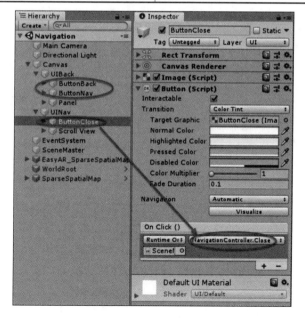

图 9-24

9.6.2 修改导航脚本

1. 添加枚举

添加一个枚举，用于记录当前状态。因为只在导航脚本中使用，所以可以不新建脚本，直接添加在导航脚本后面，与 NavigationController 对象同一级。

```
public enum NavStatus
{
    //等待
    wating,
    //定位
    localized,
    //导航
    navigation
}
```

2. 添加属性

添加稀疏空间地图属性，显示文本、导航按钮及其状态。在 Start 方法中不直接烘焙，默认关闭导航界面，设置状态。

```
public ARSession session;
public SparseSpatialMapWorkerFrameFilter mapWorker;
public SparseSpatialMapController map;
public Text textShow;
public Button btnNav;
private NavStatus navStatus;
void Start()
{
    game = FindObjectOfType<GameController>();
    lineRenderer.gameObject.SetActive(false);
    LoadObjects();
    LoadTarget();
    LoadRoad();
    Close();
    navStatus = NavStatus.wating;
    btnNav.interactable = false;
}
```

3. 添加修改导航方法

可以先将模型场景的导航方法复制过来，大的框架是一样的。

```
void Start()
{
    ...
    LoadMap();
}
```

```
private void LoadMap()
{
    //设置地图
    map.MapManagerSource.ID = game.GetMapID();
    map.MapManagerSource.Name = game.GetMapName();
    //地图获取反馈
    map.MapLoad += (map, status, error) =>
    {
        if (status)
        {
            textShow.text = "地图加载成功。";
        }
        else
        {
            textShow.text = "地图加载失败：" + error;
        }
    };
    //定位成功事件
    map.MapLocalized += () =>
    {
        textShow.text = "稀疏空间定位成功。";
    };
    //停止定位事件
    map.MapStopLocalize += () =>
    {
        textShow.text = "停止稀疏空间定位。";
    };
    textShow.text = "开始本地化稀疏空间。";
    mapWorker.Localizer.startLocalization();    //本地化地图
}
```

在定位成功和停止定位的事件中添加判断，这里用了 switch 语句。一开始定位成功后，显示导航界面。如果在等待选择导航目的地的时候定位消失就关闭导航界面。

```
//定位成功事件
map.MapLocalized += () =>
{
    textShow.text = "稀疏空间定位成功。";
    switch (navStatus)
    {
        case NavStatus.wating:
            navStatus = NavStatus.localized;
            btnNav.interactable = true;
            SetStaticObject();
            ShowNav();
            break;
        default:
```

```
            break;
    }
};
//停止定位事件
map.MapStopLocalize += () =>
{
    textShow.text = "停止稀疏空间定位。";
    switch (navStatus)
    {
        case NavStatus.localized:
            navStatus = NavStatus.localized;
            btnNav.interactable = false;
            Close();
            break;
        default:
            break;
    }
};
```

修改按钮点击事件,当按钮点击以后进入导航状态,显示导航线,烘焙路径。

```
public void SelectButtonClicked(Transform btn)
{
    navStatus = NavStatus.navigation;
    lineRenderer.gameObject.SetActive(true);
    BakePath();
    ...
}
```

根据设置,定位成功事件会触发多次,导航过程中如果触发定位成功则重新烘焙路径、重新显示导航线。为了避免重复,显示导航线之前先停止原有的重复运行。

导航过程中如果定位停止,则不再更新导航线。

```
//定位成功事件
map.MapLocalized += () =>
{
    textShow.text = "稀疏空间定位成功。";
    switch (navStatus)
    {
        case NavStatus.wating:
            ...
        case NavStatus.navigation:
            CancelInvoke("DisplayPath");
            BakePath();
            InvokeRepeating("DisplayPath", 0, refresh);
            break;
        default:
            break;
```

```
            }
        };
        //停止定位事件
        map.MapStopLocalize += () =>
        {
            textShow.text = "停止稀疏空间定位。";
            switch (navStatus)
            {
                case NavStatus.localized:
                    ...
                case NavStatus.navigation:
                    CancelInvoke("DisplayPath");
                    break;
                default:
                    break;
            }
        };
```

为脚本赋值，如图 9-25 所示。游戏对象对应属性如表 9-11 所示。

表 9-11　为脚本属性赋值

游戏对象	属　　性
ButtonNav	Btn Nav
Text	Text Show
EasyAR_SparseSpatialMapWorker	Session
SparseSpatialMapWorker	Map Worker
SparseSpatialMap	Map

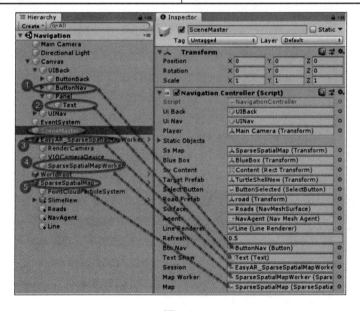

图 9-25

完整的代码如下：

```csharp
using UnityEngine;
using UnityEngine.UI;
using UnityEngine.AI;
using easyar;

namespace Kanamori
{
    public class NavigationController : MonoBehaviour
    {
        public GameObject uiBack;
        public GameObject uiNav;
        private GameController game;
        public Transform player;
        public ShowChildObject[] staticObjects;
        public Transform ssMap;
        public Transform blueBox;
        public Transform svContent;
        public Transform targetPrefab;
        public SelectButton selectButton;
        public Transform roadPrefab;
        public NavMeshSurface surface;
        public NavMeshAgent agent;
        private NavMeshPath path;
        private Transform target;
        public LineRenderer lineRenderer;
        public float refresh;
        public Button btnNav;
        public Text textShow;
        public SparseSpatialMapWorkerFrameFilter mapWorker;
        public SparseSpatialMapController map;
        private NavStatus navStatus;
        void Start()
        {
            game = FindObjectOfType<GameController>();
            lineRenderer.gameObject.SetActive(false);
            LoadObjects();
            LoadTarget();
            LoadRoad();
            Close();
            navStatus = NavStatus.wating;
            btnNav.interactable = false;
            LoadMap();
        }
        private void LoadMap()
        {
```

```csharp
//设置地图
map.MapManagerSource.ID = game.GetMapID();
map.MapManagerSource.Name = game.GetMapName();
//地图获取反馈
map.MapLoad += (map, status, error) =>
{
    if (status)
    {
        textShow.text = "地图加载成功。";
    }
    else
    {
        textShow.text = "地图加载失败：" + error;
    }
};
//定位成功事件
map.MapLocalized += () =>
{
    textShow.text = "稀疏空间定位成功。";
    switch (navStatus)
    {
        case NavStatus.wating:
            navStatus = NavStatus.localized;
            btnNav.interactable = true;
            SetStaticObject();
            ShowNav();
            break;
        case NavStatus.navigation:
            CancelInvoke("DisplayPath");
            BakePath();
            InvokeRepeating("DisplayPath", 0, refresh);
            break;
        default:
            break;
    }
};
//停止定位事件
map.MapStopLocalize += () =>
{
    textShow.text = "停止稀疏空间定位。";
    switch (navStatus)
    {
        case NavStatus.localized:
            navStatus = NavStatus.localized;
            btnNav.interactable = false;
            Close();
            break;
```

```csharp
                    case NavStatus.navigation:
                        CancelInvoke("DisplayPath");
                        break;
                    default:
                        break;
                }
            };
            textShow.text = "开始本地化稀疏空间。";
            mapWorker.Localizer.startLocalization();    //本地化地图
        }
        private void SetStaticObject()
        {
            foreach (var item in staticObjects)
            {
                item.SetVisible((item.transform.position - player.position).magnitude <= 2);
            }
        }
        public void Close()
        {
            uiBack.SetActive(true);
            uiNav.SetActive(false);
        }
        public void ShowNav()
        {
            uiBack.SetActive(false);
            uiNav.SetActive(true);
        }
        public void Back()
        {
            if (game)
            {
                CancelInvoke("DisplayPath");
                game.BackMenu();
            }
        }
        private void LoadObjects()
        {
            if (game)
            {
                var list = game.LoadDynamicObject();
                foreach (var item in list)
                {
                    var dynamicObject = JsonUtility.FromJson<DynamicObject>(item);
                    var tf = Instantiate(blueBox, ssMap);
                    tf.localPosition = dynamicObject.position;
                    tf.localEulerAngles = dynamicObject.rotation;
```

```csharp
                tf.localScale = dynamicObject.scale;
                var obj = tf.GetComponent<ShowSelfObject>();
                obj.SetVisible((tf.position - player.position).magnitude <= 2);
            }
        }
    }
    private void LoadTarget()
    {
        if (game)
        {
            var list = game.LoadKeyPoint();
            foreach (var item in list)
            {
                KeyPoint point = JsonUtility.FromJson<KeyPoint>(item);

                if (point.pointType == 0)
                {
                    var target = Instantiate(targetPrefab, ssMap);
                    target.localPosition = point.position;
                    target.GetComponent<ShowChildObject>().SetVisible(false);
                    var btn = Instantiate(selectButton, svContent);
                    btn.GetComponentInChildren<Text>().text = point.name;
                    btn.target = target;
                }
            }
        }
    }
    private void LoadRoad()
    {
        if (game)
        {
            var list = game.LoadRoad();
            foreach (var item in list)
            {
                var road = JsonUtility.FromJson<Road>(item);
                var tfRoad = Instantiate(roadPrefab, ssMap.Find("Roads"));

                tfRoad.localPosition = (road.startPosition + road.endPosition) / 2;
                tfRoad.LookAt(road.startPosition);
                tfRoad.localScale = new Vector3(0.02f, 1f, (road.endPosition - road.startPosition).magnitude * 0.1f + 0.2f);
            }
        }
    }
    private void BakePath()
    {
        agent.enabled = false;
```

```csharp
            surface.BuildNavMesh();
            path = new NavMeshPath();
        }
        private void DisplayPath()
        {
            agent.transform.position = player.position;
            agent.enabled = true;
            agent.CalculatePath(target.position, path);
            lineRenderer.positionCount = path.corners.Length;
            lineRenderer.SetPositions(path.corners);
            agent.enabled = false;
        }
        public void SelectButtonClicked(Transform btn)
        {
            navStatus = NavStatus.navigation;
            lineRenderer.gameObject.SetActive(true);
            BakePath();
            CancelInvoke("DisplayPath");
            target = btn.GetComponent<SelectButton>().target;
            InvokeRepeating("DisplayPath", 0, refresh);
            Close();
        }
    }
    public enum NavStatus
    {
        wating,
        localized,
        navigation
    }
}
```

第 10 章
◀ 调 试 发 布 ▶

10.1 发布调试应用建立地图

1. 发布调试应用

单击菜单 File→Build Settings，打开 Build Settings 窗口。将所有场景拖到 Scenes In Build 中，确保 Menu 场景是第一个场景。单击 Build 按钮发布一个测试用的 apk，如图 10-1 所示。

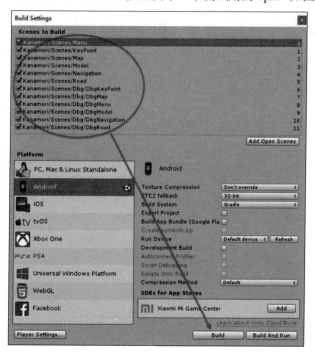

图 10-1

2. 建立地图

在菜单场景中，单击"添加地图"按钮的时候必须注意，因为静态的模型位置是由稀疏空间地图的位置决定的，这里设置的是在屏幕正前方 2 米左右的位置，所以需要把手机对准想要的方向才能单击按钮。

这里对准屏幕左边，如图 10-2 所示。扫描整个空间并保存稀疏空间地图，如图 10-3 所示。

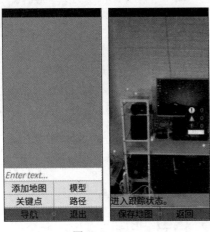
图 10-2

图 10-3

10.2 调试错误修改

1. 发现错误

如果在添加模型的场景出错,只添加了 1 个模型,保存提示却是 2 个模型,会自动在稀疏空间地图的原点生成 1 个模型。

尝试后发现,真实的稀疏空间地图下面有一个显示特征点的游戏对象,在方便调试用的场景中,模拟的稀疏空间地图下没有这个游戏对象。

2. 修改问题

打开 DbgModel 场景,在 FakeSSMap 游戏对象下添加一个空的游戏对象,并命名为 PointCloudParticleSystem,如图 10-4 所示。

选择在调试场景修改的好处是一旦保存就能准确定位到错误的位置,而且不需要浪费时间在打包的过程中,如图 10-5 所示。

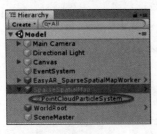
图 10-4

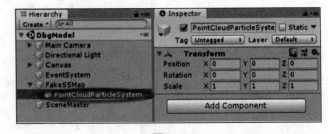
图 10-5

修改 DbgModelController 脚本的保存方法。

```
public void Save()
{
    string[] jsons = new string[ssMap.childCount - 1];
    for (int i = 0; i < ssMap.childCount; i++)
    {
```

```
            if (ssMap.GetChild(i).name != "PointCloudParticleSystem")
            {
                DynamicObject dynamicObject = new DynamicObject();
                dynamicObject.position = ssMap.GetChild(i).localPosition;
                dynamicObject.rotation = ssMap.GetChild(i).localEulerAngles;
                dynamicObject.scale = ssMap.GetChild(i).localScale;
                jsons[i - 1] = JsonUtility.ToJson(dynamicObject);
            }
        }
        ...
    }
```

从 DbgMenu 场景开始运行，确定正确以后将修改复制到 ModelController，再重新发布。

3．添加模型

添加一个模型并保存，如图 10-6 所示。

图 10-6

10.3 其他场景设置

1．关键点设置

将打印好的图像放在地上，当平面图像跟踪发现图像后会出现红色方块，点击红色方块，添加关键点，如图 10-7 所示。

这里添加了 5 个关键点，有 3 个是目的地。

2．路径场景设置

添加路径，如图 10-8 所示。将关键点连接起来，如图 10-9 所示。

图 10-7　　　　　　　　图 10-8　　　　　　　　图 10-9

10.4 最终导航

1. 定位和界面显示

进入导航场景以后,定位成功就会跳出导航菜单,如图 10-10 所示。点击目的地后,就会出现蓝色导航线,如图 10-11 所示(图中导航线为蓝色)。

2. 静态模型检查

静态的模型位置基本上是在预计的位置,只是稍微高了些,毕竟开始建立场景的时候没有严格测量距离,如图 10-12 所示。

图 10-10

图 10-11

图 10-12

3. 动态添加模型检查

当靠近到动态添加的蓝色方块时会自动显示,位置基本正确,如图 10-13 所示。当接近目的地的时候也会出现模型,如图 10-14 所示。

图 10-13

图 10-14

10.5　最终清理

打开 Menu 场景，删除 IngameDebugConsole 游戏对象，如图 10-15 所示。

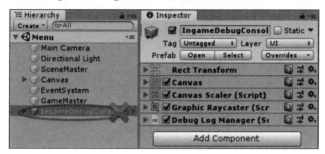

图 10-15

删除调试用的场景，再次打包，如图 10-16 所示。

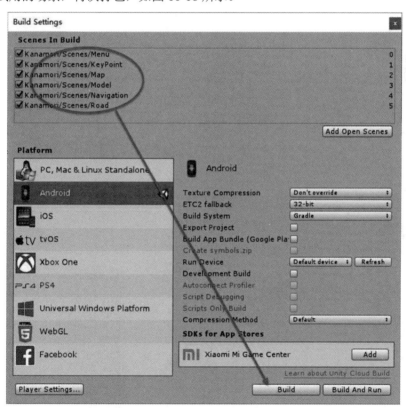

图 10-16

这样，整个项目就完成了。